Geoenvironmental Ecology, Biodiversity and Climate Change

ABOUT THE BOOK

Environmental geography is the branch of geography that describes the spatial aspects of interactions between humans and he natural world. It requires an understanding of the dynamics of geology, meteorology, hydrology, biogeography, ecology and geomorphology as well as the ways in which human societies conceptualize the environment. Geography is, in the broadest sense, an education for life and for living. Learning through geography – whether gained through formal learning or experientially through travel, fieldwork and expeditions – helps us all to be more socially and environmentally sensitive, informed and responsible citizens and employees. Ecology is the study of the interactions between organisms and their environment. Ecologists might investigate the relationship between a population of organisms and some physical characteristic of their environment, such as concentration of a chemical; or they might investigate the interaction between two populations of different organisms through some symbiotic or competitive relationship. This book is a highly informative source on global warming and climate change--issues that are perhaps the greatest threat to this planet. It will be a valuable reference tool for environmental scientists, activists and scholars in the field.

ABOUT THE AUTHOR

Naidoo Yaqub, finished her master in technology and currently pursing her PhD in environmental sciences from Institute of technology, Carlow, Irland. She is keen enthusiast of Environmental protection and Management. An alumunus of Indo-German centre for susutainibilty which is essential a collaborative interface to gather scientists in the feild of environmental sciences to come together and work for the comman cause which is environment protection.

Geoenvironmental Ecology, Biodiversity and Climate Change

Naidoo Yaqub

WESTBURY PUBLISHING LTD.
ENGLAND (UNITED KINGDOM)

Geoenvironmental Ecology, Biodiversity and Climate Change
Edited by: Naidoo Yaqub
ISBN: 978-1-913806-43-9 (Hardback)

Published by **Westbury Publishing Ltd.**
Address: 6-7, St. John Street, Mansfield,
Nottinghamshire, England, NG18 1QH
United Kingdom
Email: - info@westburypublishing.com
Website: - www.westburypublishing.com

British Library Cataloguing in Publication Data:
A catalogue record for this book is available from the British Library.

For more information regarding Westbury Publishing Ltd and its products, please visit the publisher's website- **www.westburypublishing.com**

Preface

Environmental geography is the branch of geography that describes the spatial aspects of interactions between humans and he natural world. It requires an understanding of the dynamics of geology, meteorology, hydrology, biogeography, ecology and geomorphology as well as the ways in which human societies conceptualize the environment. As we begin the 21st century, environmental thinkers are divided along a sharp fault line. There are the doomsayers who predict the collapse of the global ecosystem. There are the technological optimists who believe that we can feed 12 billion people and solve all our problems with science and technology. I do not believe that either of these extremes makes sense. There is a middle road based on science and logic, the combination of which is sometimes referred to as common sense. There are real problems and there is much we can do to improve the state of the environment.

Environmental science is the study of the interactions among the physical, chemical and biological components of the environment; with a focus on pollution and degradation of the environment related to human activities; and the impact on biodiversity and sustainability from local and global development. It is inherently an interdisciplinary field that draws upon not only its core scientific areas, but also applies knowledge from other non-scientific studies such as economics, law and social sciences. Physics is used to understand the flux of material and energy interaction and construct mathematical models of environmental phenomena. Chemistry is applied to understand the molecular interactions in natural systems. Biology is fundamental to describing the effects within the plant and animal kingdoms.

Climate change mitigation is action to decrease the intensity of radiative forcing in order to reduce the potential effects of global warming. Mitigation is distinguished from adaptation to global warming, which involves acting to tolerate the effects of global warming. Most often, climate change mitigation scenarios involve reductions in the concentrations of greenhouse gases, either by reducing their sources or by increasing their sinks. Scientific consensus on global warming, together with the precautionary principle and the fear of abrupt climate change is leading to increased effort to develop new technologies and sciences and carefully manage others in an attempt to mitigate global warming. Most means of mitigation appear effective only for preventing further warming, not at reversing existing warming. The Stern Review identifies several ways of mitigating climate change. These include reducing demand for emissions-intensive goods and services, increasing efficiency gains, increasing use and development of low-carbon technologies, and reducing fossil fuel emissions.

Environmental Ecology deals extensively with the elements, components, natural resources, physical, chemical, geological and biological aspects of environment. The last thirty years have witnessed a vast expansion of both interest and knowledge in environmental sciences.

A natural disaster is the consequence of when a potential natural hazard becomes a physical event (e.g. volcanic eruption, earthquake, landslide) and this interacts with human activities. Human vulnerability, caused by the lack of planning, lack of appropriate emergency management or the event being unexpected, leads to financial, structural, and human losses. The resulting loss depends on the capacity of the population to support or resist the disaster, their resilience.

This is a reference book. All the matter is just compiled and edited in nature, taken from the various sources which are in public domain.

This book is a highly informative source on global warming and climate change--issues that are perhaps the greatest threat to this planet. It will be a valuable reference tool for environmental scientists, activists and scholars in the field.

—Editor

Contents

Biodiversity and Sustainability Management

Biological diversity (or biodiversity) is the diversity of plants, animals and other living organisms in all their forms and levels of organization, and includes the diversity of genes, species and ecosystems, as well as the evolutionary and functional processes that link them. Developing a biodiversity conservation strategy that is based on a variety of management strategies for individual species is neither feasible nor effective. The impact of forest management practices on many species is unknown and certain practices that benefit some species are often detrimental to others. Recommended instead is the development of an ecosystem management approach that provides suitable habitat conditions for all native species. In this way, habitat diversity is used as a surrogate to maintain biodiversity.

At the same time, however, special efforts may be needed to protect the habitat of species known to be at risk, such as threatened, endangered, or regionally important species. Specific strategies for addressing these species are outlined in the *Managing Identified Wildlife Guidebook*.

Planning to maintain biodiversity should occur at a variety of levels all of which are linked hierarchically: provincial (such as the provincial biodiversity strategy), regional (such as the planning being carried out by the Commission on Resources and the Environment), subregional (such as the planning being carried out through Land and Resource Management Planning), landscape, and stand. This book applies to two of those levels: landscape and stand.

The biodiversity management approach described here is based on ecological principles and will be refined over time as new knowledge is obtained and management practices evolve. The underlying assumption of this approach is that all native species and ecological processes are more likely to be maintained if managed forests are made to resemble those forests created by the activities of natural disturbance agents such as fire, wind, insects, and disease. It has been these natural ecological processes, along with burning by aboriginal peoples, that have determined the composition, size, age, and distribution of forest types on the landscape, as well as the structural characteristics of forest stands.

Principles and assumptions

- The more that managed forests resemble the forests that were established from natural disturbances, the greater the probability that all native species and ecological processes will be maintained.
- The habitat needs of most forest and range organisms can be provided for by:
  - maintaining a variety of patch sizes, several stages, and forest stand attributes and structures across a variety of ecosystems and landscapes
  - maintaining connectivity of ecosystems in such a manner as to ensure the continued dispersal and movement of forest-and range-dwelling organisms across the landscape
  - providing forested areas of sufficient size to maintain forest interior habitat conditions and to prevent the formation of excessive edge habitat.
- To sustain genetic and functional diversity, a broad geographic distribution of ecosystems and species must be maintained within forest and range lands.
- Management for biodiversity must be flexible and adaptive. This guidebook provides recommendations rather than specific prescriptions for managing biodiversity. Success in meeting the intent of these recommendations depends on the innovativeness and creativity of land managers.
- Not all elements of biodiversity can be—or need to be—maintained on every hectare. The intent is to maintain in perpetuity all native species across their historic ranges.
- Management for biodiversity should be applied within landscapes regardless of administrative boundaries. Where natural landscapes have been administratively divided, management agencies and licensees should develop a biodiversity plan together. Landscape units are the basis on which the success of biodiversity management will be evaluated.
- The conservation of biodiversity depends on a coordinated strategy that includes:

  - a system of protected areas at the regional scale
  - provision for a variety of habitats at the landscape scale
  - management practices that provide important ecosystem attributes at the stand scale.
- Intensive forestry and other resource development within managed landscapes can be compatible with the maintenance of biological diversity.
- Where past forest management practices have resulted in forest conditions that prevent biodiversity objectives from being achieved, biodiversity can be restored over time by managing the forest to create—or recover—the required ecosystem elements.

BIODIVERSITY

Biodiversity is the variation of life forms within a given ecosystem, biome, or for the entire Earth. Biodiversity is often used as a measure of the health of biological systems. The biodiversity found on Earth today consists of many millions of distinct biological species, which is the product of nearly 3.5 billion years of evolution.

Evolution and Meaning

Biodiversity is a portmanteau word, from biology and diversity, originating from and used interchangeably with "biological diversity." This term was used first by wildlife scientist and conservationist Raymond F. Dasmann in a lay book advocating nature conservation. It was not widely adopted for more than a decade, when in the 1980s it and "biodiversity" came into common usage in science and environmental policy. Use of the term by Thomas Lovejoy in the Forward to the book credited with launching the field of conservation biology introduced the term along with "conservation biology" to the scientific community. Until then the term "natural diversity" was used in conservation science circles, including by The Science Division of The Nature Conservancy in an important 1975 study, "The Preservation of Natural Diversity." By the early 1980s TNC's Science programme and its head Robert E. Jenkins, Lovejoy, and other leading conservation scientists at the time in America advocated the use of "biological diversity" to embrace the object of biological conservation.

Its contracted form biodiversity *may have been coined by W.G. Rosen in 1985 while planning the* National Forum on Biological Diversity *organized by the National Research Council (NRC) which was to be held in 1986, and first misha appeared in a publication in 1988 when entomologist E. O. Wilson used it as the title of the proceedings of that forum.*

Since this period both terms and the concept have achieved widespread use among biologists, environmentalists, political leaders, and concerned citizens worldwide. It is generally used to equate to a concern for the natural environment and nature conservation. This use has coincided with the expansion of concern over extinction observed in the last decades of the 20th century.

A similar concept in use in the United States, besides natural diversity, is the term "natural heritage." It pre-dates both terms though it is a less scientific term and more easily comprehended in some ways by the wider audience interested in conservation. "Natural Heritage" was used when Jimmy Carter set up the Georgia Heritage Trust while he was governor of Georgia; Carter's trust dealt with both natural and cultural heritage. It would appear that Carter picked the term up from Lyndon Johnson, who used it in a 1966 Message to Congress. "Natural Heritage" was picked up by the Science Division of the US Nature Conservancy when, under Jenkins, it launched in 1974 the network of State Natural Heritage Programmes. When this network was extended outside the USA, the term "Conservation Data Center" was suggested by Guillermo Mann and came to be preferred.

Definitions

Biologists most often define "biological diversity" or "biodiversity" as the "totality of genes, species, and ecosystems of a region". An advantage of this definition is that it seems to describe most circumstances and present a unified view of the traditional three levels at which biological variety has been identified:

- genetic diversity
- species diversity
- ecosystem diversity

This multilevel conception is consistent with the early use of "biological diversity" in Washington. D.C. and international conservation organizations in the late 1960s through 1970's, by Raymond F. Dasmann who apparently coined the term and Thomas E. Love joy who later introduced it to the wider conservation and science communities. An explicit definition consistent with this interpretation was first given in a paper by Bruce A. Wilcox commissioned by the International Union for the Conservation of Nature and Natural Resources (IUCN) for the 1982 World National Parks Conference in Bali The definition Wilcox gave is "Biological diversity is the variety of life forms...at all levels of biological systems (*i.e.*, molecular, organismic, population, species and ecosystem)..." Subsequently, the 1992 United Nations Earth Summit in Rio de Janeiro defined "biological diversity" as "the variability among living organisms from all sources, including, 'inter alia', terrestrial, marine, and other aquatic ecosystems, and the ecological complexes of which they are part: this includes diversity within species, between species and of ecosystems". This is, in fact, the closest thing to a single legally accepted definition of biodiversity, since it is the definition adopted by the United Nations Convention on Biological Diversity. The current textbook definition of "biodiversity" is "variation of life at all levels of biological organization".

If the gene is the fundamental unit of natural selection, according to E. O. Wilson, the real biodiversity is genetic diversity. For geneticists, *biodiversity* is the diversity of genes and organisms. They study processes such as mutations, gene exchanges, and genome dynamics that occur at the DNA level and generate evolution. Consistent with this, along with the above definition the Wilcox paper stated "genes are the ultimate source of biological organization at all levels of biological systems..."

BIODIVERSITY OBJECTIVES

Once landscape units have been delineated and the biodiversity emphasis option determined, objectives for maintaining biodiversity within the units may be set for the following characteristics:

- several stage distribution
- temporal and spatial distribution of the cut and leave areas "patch size distribution"
- old several retention and representativeness
- landscape connectivity
- stand structure
- species composition.

British Columbia has tremendous ecological variation. Although some general forest and range management practices can broadly accommodate the needs of all ecosystems, more often a variety of practices is needed to respond to the different natural disturbance regimes under which ecosystems have evolved. For the purpose of setting biodiversity objectives, five natural disturbance types (NDTs) are recognized as occurring in British Columbia. They are:

NDT1 – Ecosystems with rare stand-initiating events

NDT2 – Ecosystems with infrequent stand-initiating events

NDT3 – Ecosystems with frequent stand-initiating events

NDT4 – Ecosystems with frequent stand-maintaining fires

NDT5 – Alpine Tundra and Subalpine Parkland ecosystems

These disturbance types characterize areas with different natural disturbance regimes. Stand-initiating disturbances are those processes that largely terminate the existing forest stand and initiate secondary succession in order to produce a new stand. The disturbance agents are mostly wildfires, windstorms and, to a lesser extent, insects and landslides. The stand-maintaining disturbances—such as the understorey surface fires that occur in the Interior Douglas-fir and Ponderosa Pine biogeoclimatic zones—serve to keep successional processes stable.

The following descriptions of the five NDTs include the biogeoclimatic zones, subzones, and variants that fall under each disturbance regime. As additional information becomes available, the number of NDTs and the biogeoclimatic units within them may be revised and refined.

Recommended procedure for establishing landscape unit biodiversity objectives

- Confirm the biodiversity emphasis option for the landscape unit.
- Map the NDTs within the landscape unit, based on the biogeoclimatic subzones and variants present in the unit.
- Use the NDT recommendations to establish the biodiversity objectives for the landscape unit.
- Where a landscape unit consists entirely of one NDT, the biodiversity objectives for the landscape unit should be those of the NDT. However, where a landscape unit consists of more than one disturbance type, biodiversity objectives for the landscape unit should be developed for each of the NDTs present. In areas that are transitional between two NDTs, the management practices should be modified to reflect that transition.
- Where a disturbance type accounts for only a small, isolated part of a landscape unit, meeting the landscape level objectives for the small area may not be possible. In such cases include the small area in an adjacent disturbance type.

IMPACT ON BIODIVERSITY

Biodiversity existing in dry lands and other habitats underpin ecosystem services that vital for livelihoods of millions of people in Africa. It is the foundation for sustainable development in the region and globally. The dry areas of the world are the origin of a large number of globally important cereals and food legumes, such as barley, wheat, faba beans and lentils. Four hundred million people, two thirds of sub-Sahara African population, rely on forest goods and services for their livelihood. Drought, land degradation and desertification have had serious impact on the richness and diversity of Africa Diversity. These factors remain some of the most serious threats to the management, sustainable use and equitable sharing of benefits of biodiversity. The projected devastating impacts of climate change in the region including exacerbating these factors will escalate biodiversity degradation and loss associated with drought, land degradation and desertification.

These factors affect biodiversity directly and indirectly. Onsite impacts include habitat and species degradation and loss, leading to overall loss of economic and biological productivity. For instance on rangelands, overgrasing not only reduces the overall protective soil cover and increases soil erosion, but also leads to a long-term change in the composition of the vegetation. Plant biodiversity will change over time, unpalatable species will dominate, and total biomass production will be reduced. These in turn trigger and contribute to indirect or offsite impacts. Soil erosion will contribute to denudation and pollution of wetlands and water bodies. As biological and economic productivity deteriorates, communities are forced migrate to other areas or engage in other coping activities that too contribute biodiversity degradation.

According to the Africa Environment Outlook II, approximately half of Africa's terrestrial eco-regions have lost more than 50 per cent of their area to cultivation, degradation or urbanisation. It also states that some ecoregions such as the Mandara Plateau mosaic, Cross-Niger transition forests, Jos Plateau forest-grassland mosaic, and Nigerian lowland forests have gone more than more than a 95 percent transformation.

Nine other eco-regions have lost more than 80 per cent of their habitat, including the species-rich lowland Fynbos and Renosterveld and the forests and grasslands of the Ethiopian Highlands; the Mediterranean woodlands and forests have lost more than 75 per cent of their original habitat, and the few remaining blocks of habitat are highly fragmented.In the sand dune areas of countries such as Mauritania, Mali, Niger, Nigeria and Senegal major river basins siltation processes accumulate debris and materials that engulf natural vegetation, such as the *Acacia nilotica* riparian forests.

Soil erosion contributes to moving the seed capital of the ground, uprooting grassy as well as woody species, and in accumulation areas it smothers valuable species.

In West Africa the movement of people south towards subhumid to humid tropical areas has 38. resulted into loss of primary forests and woodlands, repeated logging of the secondary vegetation, and depletion of a number of species (UNEP 2006). More diffuse degradation of land resources also occurs in the arid and sub-humid parts. These include the extraction of tree resources outside forests for charcoal making (about 150 million tonnes/year from the savannah and woodland areas), and the use of high-value woods. Most affected are the *Meliacaea family* (*Khaya species*), *Pterocarpus erinaceus,* and *Dalbergia melanoxylon.* There is mounting evidence to show that drought and desertification as exacerbated by climate change will have devastating impacts on habitats and species in the region. For example shifts in rainfall patterns could affect the fynbos and karoo in southern Africa by altering the fire regime critical for their regeneration. Decreasing run off could impact wetland ecosystems such as the Okavango Delta and the Sudd area.

Impact on Migration

The effects of desertification extend beyond the affected dryland areas. As the level of vulnerability due to the combined impacts of desertification and socio-economic susceptibility increase, the greater the probability of human migration. Desertification is displacing big population of people and forcing them to leave their homes and lands in search of better livelihoods. Desertification and drought related migration takes many forms the majority occurring as internal migrations (Nanyunja, 2004), that is, displacements of populations within national boundaries (Mora and Taylor, 2006; Lein, 2000; Zaman, 1991). At greatest risk are those at the low end of the socio-economic spectrum, both in developed and developing regions.

In developing regions, the poorest inhabitants are often forced to live on marginal land outside urban areas or coastal zones , potentially prone to desertification. Migration is often a coping mechanism, with little faith in finding permanent residence (Haque and Zaman, 1989; Mutton and Haque, 2004; Zaman, 1991). Availability of natural resources for example prompts pastoralists along the borders of Ethiopia, Kenya and Uganda to migrate away from areas of dwindling resources; thus raising competition over finite resources with incidence of conflict increasing when these individuals move into areas of crop growing communities. It is estimated that 135 million people - the combined populations of France and Germany - are at risk of being displaced by desertification. The problem appears to be most severe in sub-Saharan Africa, the Sahel and the Horn of Africa. Some 60 million are estimated to eventually move from the desertified areas of sub-Saharan Africa towards Northern Africa and Europe by the year 2020.

It has migrated at least once to neighbouring African countries (96 percent) or to Europe (2.7 percent). In Burkina Faso, desertification can be identified as the cause of 60 percent of the swelling of main urban centres. In Kenya one of the consequences of desertification is a constant flow of rural poor to Nairobi. The population of Nairobi has grown by 800 percent from 350,000 in 1963 to 2,818,000 in 2005. Migration will exert stress on the poor and limited public infrastructure in urban areas and may exacerbate conflicts already witnessed in the region as result of scarcity of grasing land and water. Against this background of the devastating impact of drought and desertification, which permeates and undermines the very foundations for securing sustainable livelihoods and economic growth, poverty eradication in Africa is inextricably linked to success in combating desertification and mitigating the impacts of drought. For millions on the continent, hopes of getting out of poverty therefore hinge on efforts at national, regional and global levels to prioritise the provision of support and the implementation measures for desertification control and coping with drought.

Impact on Energy

The impacts of drought and desertification on the energy sector are felt primarily through losses in hydropower potential for electricity generation and the effects of increased runoff (and consequent siltation) on hydropower generation as demonstrated. The gravity of impacts of electricity generation is further demonstrated by the case of Ghana, where for the first half of 2007 (and it was projected to continue for the year), the water level at the Akosombo dam had fallen below the minimum level of 240 feet. This led to reduction in hydro-electricity generation and hence load shedding of electricity in the whole country. Energy impacts are also experienced through changes in the growth rates of trees on which a vast majority of the people in the region rely for fuel wood. Due to the limited alternatives available to them and low priority accorded to meet their needs in times of scarcity, the rural areas and the urban poor bear the greatest cost of decrease in energy resources. This undermines efforts to pull these categories of people out of the poverty trap.

BIODIVERSITY EMPHASIS OPTIONS TO LANDSCAPE UNITS

Applying biodiversity emphasis options to landscape units across a subregional planning area is a key part of a biodiversity management strategy.

Biodiversity emphasis options can be assigned to landscape units in the following ways:

- Direct assignment of a biodiversity emphasis option to a landscape unit by a regional or subregional land use planning process.
- Indirect assignment of biodiversity emphasis options to landscape units by a regional or subregional land use planning process that develops broad-scale land use zonation. In such instances agency representatives will assign a specific landscape unit an emphasis option consistent with the zonation.
- In the absence of regional or subregional plans, interim biodiversity emphasis options can be assigned to a landscape unit by joint

agreement of the district manager and the designated environmental official.

If no landscape unit has been designated for an area, or an emphasis option has not been assigned for a landscape unit, the default is that the area is managed using the lower biodiversity emphasis option. To assist with the process of assigning biodiversity emphasis options to landscape units, each unit can be evaluated according to several criteria: topographic and ecosystem complexity, wildlife and fisheries species diversity, significance of key management species and social and economic considerations. Government has evaluated social and economic impacts against risk to biodiversity on a provincial basis and provided the following policy direction concerning landscape unit biodiversity emphasis assignments within subregional areas. In some cases, large portions of a landscape unit may require a different biodiversity emphasis than the rest of the landscape unit. For example, a low elevation subzone may have a different biodiversity emphasis option assigned than a high elevation subzone. Within a landscape unit, minor areas (up to thousands of hectares in size), may be managed differently from the rest of the unit.

For example, a landscape unit assigned a lower biodiversity emphasis option could contain some small protected areas that would contribute to the seral stage and patch size requirements for the landscape unit. Alternatively, stands managed for intensive timber production may be part of any landscape unit, regardless of the biodiversity emphasis option assigned the major portion. Concentration of the lower biodiversity emphasis option over extensive, contiguous areas of a subzone or subregional planning area (over several adjacent landscape units, for example) should be avoided as it will greatly impact natural levels of biodiversity in that area. Instead, landscape units with the lower biodiversity emphasis should be distributed across the subregional planning area.

BIODIVERSITY EMPHASIS OPTIONS

As natural ecosystems become increasingly modified by human activities, natural patterns of biodiversity become increasingly altered and the risk of losing native species increases. The greatest degree of disruption occurs from extreme habitat modifications such as urbanization and agriculture. Parks and protected areas, on the other hand, if appropriately managed, maintain close to natural levels of biodiversity. Managed forest lands fall between those two extremes, and can support varying levels of natural biodiversity, depending on the management practices. More natural levels of biodiversity will be maintained in managed forests if those forests are managed to mimic important characteristics of natural forest conditions.

This guidebook outlines a range of three options for emphasizing biodiversity at the landscape level. Each option is designed to provide a different level of natural biodiversity and a different risk of losing elements of natural biodiversity:

- The lower biodiversity emphasis option may be appropriate for areas where other social and economic demands, such as timber supply, are the primary management objectives. This option will provide habitat for a wide range of native species, but the pattern of natural biodiversity will be significantly altered, and the risk of some native species being unable to survive in the area will be relatively high.
- The intermediate biodiversity emphasis option is a trade-off between biodiversity conservation and timber production. Compared to the lower biodiversity emphasis option, this one will provide more natural levels of biodiversity and a reduced risk of eliminating native species from the area.
- The higher biodiversity emphasis option gives a higher priority to biodiversity conservation but would have the greatest impact on timber harvest. This option is recommended for those areas where biodiversity conservation is a high management priority.

In reality, these options are points on a continuum, and in between lie a range of options that may be selected depending on the relative priority allocated to biodiversity conservation and timber production in an area.

MICROBIAL BIODIVERSITY

The microbial world encompasses a large proportion of life forms, broadly including any organism that spends all or most of its life in microscopic form. Microorganisms therefore include viruses, single-celled algae, bacteria, archaea (a specialised form of bacteria that are important in decay processes, especially in wetlands), protozoa and fungi, and in the broadest sense, microscopic invertebrates such as nematodes and mites. During the 1990s, it has become increasingly apparent that our knowledge of biodiversity in the microbial world is largely inadequate. More than 99% of microorganisms are yet to be discovered or described. Understanding the roles of these unknown organisms is likely to be essential to understanding and monitoring biodiversity, since they occupy key positions in all ecosystems. The prokaryotes (bacteria and archaea) as a group contain roughly equivalent quantities of organic carbon as plants, and may hold up to 10 times more phosphorus and nitrogen than do plants (Whitman et al. 1998). Similarly, in terrestrial ecosystems, the biomass of fungi exceeds all other groups except vascular plants (Lal 1995). Microorganisms perform key ecosystem services such as nitrogen fixation, carbon cycling and the regulation of atmospheric gases. A recent attempt to value the world's ecosystem services arrived at a value of US$33 trillion per year. Categories of these services that are mainly, or partly provided by microorganisms (nutrient cycling, waste treatment/ degradation, atmospheric gas regulation, erosion control, soil formation, biological control and food production) amount to over 70% of this total value.

A further area where we lack adequate understanding of microbial biodiversity is in the associations between microbial life and larger organisms. Almost all macroorganisms depend on microbial symbioses to some extent. Microorganisms contribute significantly to the conservation and production of nutrients in the vertebrate gastrointestinal tract. Investigations of the microbiota associated with native Australian animals are only just beginning. Around 75% of vascular plants form mutualistic associations with mycorrhizal fungi.

Although the composition and type of plant species in a terrestrial ecosystem is a primary determinant of ecosystem productivity and sustainability, plant biodiversity may in turn be primarily regulated by the diversity of mycorrhizal fungi. Consequently, fungal diversity may indirectly control both ecosystem productivity and variability. Fungal endophytes also occur in the leaves and stems of vascular plants, and the extent and importance of this form of mutualism is receiving more attention. In Australia, 95% of ectomycorrhizal fungi are novel, with some 22 genera and 3 families being endemic. Thus, Australian fungi are likely to be as unique as our animal and plant species. A similar situation occurs in the mutualistic association between nitrogen-fixing soil bacteria and legumes, which in Australia represent about 10% of plant species. A survey of native shrubby legumes recovered 21 genomic rhizobial species, only one of which corresponded to a known species.

Because of their large biomass, extraordinary genetic diversity and their central roles in many ecosystem processes, the characterisation of microbial biodiversity should be a consideration in any biodiversity assessment. However, the systematic investigation of microbial diversity in Australia has been the subject of few studies, and methods for rapid assessment of the distribution and abundance of microorganisms are still being developed. The conservation of microbial diversity has not yet received the attention given to larger organisms, and indeed there is only one mention of the word 'microorganism' in the body of the CBD (Davison et al. 1999).

Endemism [BD Indicator 10.6]

A taxon (e.g. a species) is considered endemic to a particular area if it occurs only in that area. The proportion of vertebrate taxa in Australia that are endemic is particularly high compared to other countries, with the richness of vertebrate species being largely as a result of a remarkable variety of reptiles. Because of its size, age and geological and evolutionary isolation, over 80% of mammal, reptile and flowering plant species in Australia are also endemic. The degree of endemism in fungi, molluscs and insects is also estimated to be over 80%, and as such, the Australian continent is recognised as a centre of endemism of global significance. While the level of endemism of many taxonomic groups is still unknown, it is likely to be high because of Australia's evolutionary and geological history.

Since the concept of endemism is tied to particular areas, the identification of centres of endemism is dependent on scale. At a global level, high levels of species richness and endemism are consistent across most taxonomic groups in Australia and in most environments. Areas of high endemism also occur within regions in the Australian continent. For example, about 5% of Australia's flora occurs in the Stirling Ranges of south-west Australia, an isolated mountain range with some unique ecological characteristics. The high level of vascular plant endemism in this region is evident also at the global level. In a worldwide study, Myers identified 18 relatively small areas that are rich in endemic vascular plant species and that are experiencing relatively rapid rates of habitat modification or loss. The only region in Australia on this list is the south-west of the continent. These 18 sites contain about 50 000 endemic plant species (20% of the world's total) in about 750 000 square kilometres (0.5% of the earth's surface area).

Several different patterns of endemism have been observed within the Australian biota. Cracraft (1991) identified 14 recognisable areas in Australia with a unique assemblage of bird species. Crisp et al. (2001) analysed the distribution patterns of around 8500 vascular plant species. Twelve centres of endemism, which were all near coastal, were identified. The lack of centres of endemism in inland Australia was attributed to the selective extinction of narrow endemics driven by extreme climates during the last glacial maximum.

Major centres of both plant endemism and diversity were also examined by Crisp et al. (2001). The regions that met both these criteria were south-west Australia, the Border Ranges between New South Wales and Queensland; the Wet Tropics near Cairns; Tasmania; and the Iron-McIlwraith Range of eastern Cape York Peninsula. The last centre appears to be more significant than recognised previously, and the Adelaide-Kangaroo Island region, which was also identified as important, has previously been overlooked altogether. It is important to have identified areas where high concentrations of species occur, so that they can be sympathetically managed.

Levels of endemism are also high for marine groups such as macroalgae, with southern Australia being of major significance (Zann 1995). Levels of endemism of other marine groups are described. The 5500 km coastline from south-west Western Australia to the border between New South Wales and Victoria is particularly rich in brown algae and red algae. Of these two groups, around 57 and 75% of species are endemic to southern Australia. Recent studies of cave biota in Australia have also discovered high levels of endemism.

Another measure of the uniqueness of Australia's biodiversity is the high level of variation in species richness among communities, and the variation in commonness and rarity among species, termed mosaic diversity. High values indicate complex landscapes with many environmental gradients, and many species with roughly equal abundance. Using this measure, the diversity of species and terrestrial ecosystems in Australia exceeds that of any other continent.

Vulnerable, endangered, threatened or extinct species

The differences between taxa in the percentages of threatened and extinct species reflect the biases in taxonomic and conservation focus. Vertebrates (especially birds and mammals) and in general vascular plants receive much more attention than do invertebrates, non-vascular plants and fungi. This bias also appears in the records of other countries. It appears that the number of nationally endangered and vulnerable species has increased in several groups over the last seven years. In some instances, the numbers of species in these categories may change over time because there have been changes in the abundance or distribution of species. But in many cases, the changes are due to taxonomic revisions resulting in either the creation or loss of new species.

The status of species may also change based on new information without any underlying change in the number or distribution of individuals or in the processes affecting them. New observations result in a reassessment of area of occupancy, extent of occurrence, population size, threat status, trends in population size or other factors contributing to assessment of conservation status.

Plants

The most important change has been an increase in the number of endangered vascular plants. There were 226 endangered species on the official list in 1993, and 517 in 2001. Most of this is because the work on assessing threatened flora is continuing. Most additions are species that have never been assessed before, including some recently described taxa with very restricted distributions. However, there are some that are facing higher levels of threat than they were in 1993. The most important cause of changes in threat status that are not the result of new work or taxonomic revision over the last five years has been land clearance for urban and agricultural development.

As many as 123 species vascular plant species were presumed extinct in 1988 but the number was revised downwards subsequently, largely as a result of new information, survey work and taxonomic revision. A total of 227 vascular plant species have been considered at some time to be extinct. The first two Rare or Threatened Australian Plants (ROTAP database) lists share relatively few species with one another, or with any of the subsequent lists. The turnover in species on the list of presumed extinct plants has been substantial, varying by 10 to 40% of their constituent species, even during the 1990s when the total number of species presumed to be extinct did not change greatly. Most of the changes are due to taxonomic revision and survey work. Few reflect actual changes in status.

The number of presumed extinct vascular plants at the national level has fallen from 74 to 63 during 1993 to 2001. This has occurred as a result of taxonomic revision and rediscovery. Additional survey work is often prompted by the inclusion of species on the 'presumed extinct' list. There may be some further reduction over the next five years. There may be new additions as new species are described, data on plant abundances are revised, and surveys are conducted for rare and endangered flora.

Birds

The 2000 Action Plan for Australian Birds listed 25 bird taxa (reporting to the subspecies level) as extinct, 32 as critically endangered, 41 as endangered, 82 as vulnerable and 81 as near threatened. The remaining 1114 taxa are deemed to be least concern, including 28 introduced taxa and 95 vagrants. Of those taxa known to have been present or to have occurred regularly in Australia when Europeans settled in 1788, 1.9% are reported as extinct and a further 11.5% are considered threatened. Some 6.0% are near threatened. Since the last Action Plan in 1992, research and surveys have shown that seven taxa are less threatened than was thought but a further 56 taxa should have been listed. Other differences between the 1992 Action Plan and Garnett and Crowley (2000) are accounted for by changes to taxonomy (19 taxa), to more rigorous IUCN criteria, which better define the different categories (138 taxa) or both (11 taxa).

Using current knowledge, taxonomy and IUCN criteria, there has been a change in the status of 25 bird taxa (2.0%) over the eight years since the 1992 Bird Action Plan. For seven taxa, the conservation status can be downgraded as a result of effective conservation management: two from critically endangered to endangered, four from endangered to vulnerable and one from vulnerable to near threatened. However, the status of 18 taxa should be upgraded. Although no taxon has become extinct in the last decade, there has been a net increase of eight critically endangered taxa, with six fewer vulnerable and one more near threatened species.

Invertebrates

There are 281 species of Australian invertebrates listed on the 1996 IUCN Red List of Threatened Animals (IUCN 1996), representing fewer than 0.5% of known taxa. However, only four species are listed nationally in the EPBC Act. The first attempt to appraise an invertebrate group in Australia comprehensively is underway for butterflies. At a broader level, the draft Action Plan for Invertebrates aims to objectively assess the conservation status of invertebrates in Australia by using 25 species to illustrate the range of needs for this diverse group.

Turtles

Unlike sea snakes, turtles are a long-lived group of reptiles that are slow to reach maturity. They may breed only around five times in their lives, making them extremely vulnerable to overexploitation and habitat destruction or modification (Zann 1995). Breeding migrations may cover hundreds to thousands of kilometres and many turtles breeding in Australia may live around the islands of Papua New Guinea, the south-west Pacific Islands and Indonesia, making habitat management and enforcement difficult. The main human effects that occur while turtles are in Australian waters are: mortality of subadults and adults in prawn trawls, shark nets, drumlines and gill nets, and in collisions with high speed vessels; hunting by Indigenous communities; habitat degradation; and predation on eggs by feral animals. The effects of disease and parasites are unknown.

Large dots denotes thousands of nesting females per year, medium denotes 10 to 100 females per year and the smallest dots are <10 per year. There is a gap in knowledge for Arnhem Land and data for Western Australia has been pooled at the regional level.

Monitoring of turtles is essential to ensure that management practices are suitable to minimise or eliminate habitat degradation. In the Great Barrier Reef World Heritage Area most scientific studies of turtle populations have concentrated on Green and Loggerhead Turtles (*Chelonia mydes and Caretta caretta* respectively).

The Loggerhead is of specific concern as the number of nesting females has steadily declined since surveys began in the late 1970s. The east Australian population of Loggerhead Turtles used to represent the bulk of the South Pacific stock. If this population disappears, it will represent a highly significant loss. Because female turtles nest in the area where they were hatched, it is highly unlikely that a stock that has died out would be colonised naturally by Loggerhead Turtles elsewhere in the world.

Seabirds

About 142 species of seabirds belonging to 12 families are found in Australia and its external territories (Zann 1995). Of these, 76 species breed and spend their entire lives in the region, and 34 species are regular or occasional visitors. Problems for sea birds include illegal poaching of adults, chicks and eggs; mortality from bushfires and feral animals; incidental capture of albatrosses and other seabirds by longline fishing; clearing of habitats; decline in prey due to overfishing; and disturbances of nesting colonies by humans and low-flying aircraft. Possibly half of Australia's nesting islands are subject to one or more of these direct human threats.

The conservation status of marine species

Knowledge of the conservation status of most of Australia's marine species is very limited. Australia's first endangered marine fish, the Spotted Handfish (*Brachionichthys hirsutus*), is endemic to the lower Derwent River estuary in Tasmania. The decline in Spotted Handfish numbers has been linked to decline in suitable spawning substrate due to overall deterioration in the health of the Derwent system and the effect of the introduced Northern Pacific Seastar (*Asterias amurensis*).

Scientific interest has largely centred on the higher vertebrates such as turtles, seabirds, seals, Dugongs and whales. Microorganisms, algae, invertebrates and fish have been generally neglected. Australia is very rich in macroalgae or seaweeds. Southern Australia has over 1150 species (Zann 1995). This is greater than 50% more than any comparable region in the world.

Fish Species

Australia has an estimated 4000 to 4500 species of fish, of which around 3600 have been described (Zann 1995). About one-quarter of the species are endemic, most of which are found in the south. Although regulations governing many of the fished species have long existed in Australia, marine fish conservation is a relatively new field and the conservation status of most species is poorly known. Potentially vulnerable fish include sharks, which are slow growing, have a low reproduction rate, are highly migratory and form schools during the mating season.

Threats are commercial and sports fisheries, and shark meshing of surfing beaches. Fish species with restricted distributions are also vulnerable, particularly from loss of habitat. Broad-scale studies of the distribution of coral reef fishes, conducted by the Australian Institute of Marine Science (AIMS), show that overfishing can rapidly deplete the stocks of coral reef fishes. Surveys of the density of Coral Trout on Bramble Reef during and after the reef was closed to fishing found that the population of legal size Coral Trout fell by 57% in just two months when the reef was reopened for fishing.

Sea Snakes

Australia has about 30 of the total number of 50 species, about half of which are endemic (Zann 1995). The family of aipysurids live in coral reef waters and the family of hydrophiids live in the interreef waters of Australia's tropics. Sea snakes bear live young and have a relatively short lifespan; they reach sexual maturity in around three years, and live for some 10 years. The greatest human impact is from prawn trawling. Between 10 and 40% of sea snakes taken in trawls die once released (Barratt et al. 2001). For the past 20 years, trawled sea snakes have been used in a small leather industry. Licences limit the take of sea snakes for leather to 20 000 per year (Barratt et al. 2001).

Dugongs

The tropical Dugong (*Dugong dugon*) is the only fully herbivorous marine mammal and the only Sirenian (sea cow) to occur in Australia (Zann 1995). It is extinct or near extinct in most of its former range which extended from East Africa to South-East Asia and the Western Pacific. Northern Australia has the last significant populations (estimated to be over 80 000) in the world. Large, long-lived mammals, Dugongs become sexually mature at around nine to 17 years and calve every three to seven years, making them vulnerable to excessive mortality. Management concerns include the potential for overhunting of some Torres Strait populations, death of individuals that are accidentally caught in fish gillnets and shark nets, and loss of seagrass habitat.

Surveys of Dugongs have been undertaken across Queensland since the early 1980s. Surveys south of Cooktown have documented a distinct decline in population with the 1994 estimate being only 48% of that for 1986 to 1987.

A major mortality of animals occurred in Hervey Bay (Qld) in 1992 following the die-off of seagrasses. The Dugong is listed by the IUCN as 'vulnerable to extinction' and is a listed marine species under the EPBC Act. A population of around 10 000 animals at Shark Bay in Western Australia is considered of major significance for the species.

Seals

Australia's seals were overhunted in the 1800s. They are now fully protected and some populations appear to be increasing, although marked long-term declines have been reported for other populations such as the Southern Elephant Seal (*Mirounga leonina*) on Macquarie Island which have been monitored for the last 50 years. In this instance, the cause of the decrease is unknown, although it is likely to be related to increased competition for food supplies. An Action Plan has also been developed to encourage the long-term viability of seals. Major human threats include entanglement in fishing nets and ocean litter, oil pollution and disturbances by visitors. Fur seals are still occasionally illegally killed for lobster bait, and around fish farms for 'stealing' fish. Development of predator-resistant cages has reduced the latter problem. Entanglement in nets and plastic box straps remains a major threat. About 2% of seals at haul out or resting sites in Tasmania are entangled in net fragments and other plastic litter at any time (Barratt et al. 2001). A significant number of more badly tangled seals drown before reaching haul out sites. In 1990, an oil spill in Western Australia affected a number of New Zealand Fur Seal pups (Zann 1995).

Whales and Dolphins

Gillnets, shark nets set off bathing beaches, discarded fishing nets, bioaccumulation of toxins and ingestion of plastic litter are considered threats to cetaceans (whales and dolphins) within Australia (Zann 1995). During the 1980s, almost 14 000 dolphins were drowned in Taiwanese shark gillnets off northern Australia but this fishery is now closed. The use of long driftnets (sometimes referred to as the 'walls of death'), which caused substantial mortalities of cetaceans, is now banned under the Convention for the Prohibition of Fishing with Long Driftnets in the South Pacific and the United Nations global moratorium on their use. However, many cetaceans are still caught in protective shark nets off bathing beaches. For example, around 520 dolphins were caught in shark nets off the coast of Queensland between 1967 and 1988 (Zann 1995).

During the period of industrial whaling (1948-1962), the estimated numbers of Humpback Whales (*Megaptera novaeangliae*) fell dramatically. Whaling has been replaced by the new industry of whale watching, which can bring significant incomes to local economies. For example, in Victoria, Warrnambool's land-based whale-watching industry is estimated to contribute $17 million annually to the region. Because of concerns that boats, aircraft and divers may affect whale behaviour, regulations govern the distances that observers may approach whales. Increases in the estimated number of whale species such as the Humpback Whale in eastern Australian waters suggest that habitat conditions are sound. Numbers have shown a steady increase since regular monitoring began in 1981.

ADAPTIVE MANAGEMENT OF AGRICULTURAL BIODIVERSITY

Spatial and Temporal Variation in Agricultural Biodiversity: Some Management Implications

- Crop varieties planted out experience rapid changes in environmental conditions, both above and below ground. For example, the physico-chemical and biological characteristics of soils are rarely homogenous within a single plot, let alone between plots. The intense selective pressures associated with this kind of microgeographical variation calls for a fine grained approach to agricultural biodiversity management that hinges on participatory plant breeding and decentralised seed multiplication. This adaptive strategy is generally advocated for resource poor farming systems in marginal, risk prone environments. However, it may be increasingly relevant for high input situations where agricultural diversification is used to solve production problems induced by genetic uniformity (e.g. pest outbreaks) or to exploit new market opportunities (economic niche for local or regional products).
- The abundance of insect pests and their predators is known to vary enormously within and between fields, even in the more intensively managed systems. In high input irrigated rice farms, 100 fold differences in abundance of planthopper populations are commonly observed on rice plants grown a few metres apart. Huge variations in insect abundance also exist at larger spatial scales and all are marked by dynamic change over time. This implies that highly differentiated pest management approaches are needed to monitor and control pests effectively and economically. The FAO-Government program on IPM is a clear demonstration of the advantages of such local adaptive management of pests and their predators in irrigated rice in Asia. Farmer Field Schools have been a major innovation for the local

adaptive management of agricultural biodiversity by developing the capacity of farmers to develop site-specific crop protection solutions.

Actions

- Carry out administrative tasks, land use planning, agricultural research and development as near to the level of actual users of resources or beneficiaries of administration as is compatible with efficiency and accountability
- Ensure flexibility and diversity in institutional and organisational design to enable government administration and services to track appropriately the dynamic changes which occur in agroecosystems and the functions of agricultural biodiversity at different time and spatial scales.
- Educate policy makers, professionals and the public (including the bearers of local knowledge) about the value of local and indigenous knowledge and management systems in sustaining agricultural biodiversity and its many functions
- Strengthen local groups and institutions by devolving resources and removing administrative or legal hurdles to local planning and action. Support the development of local institutions for common property resources and the equitable sharing of benefits from their utilisation
- Identify and support a mediator for conflict resolution and an arbiter of last resort; guaranteeing a level legal playing field and equality of advocacy in disputes, both within and between local groups as well as between local groups and powerful external interests. Of particular importance in this connection are government policies and actions that explicitly prevent discrimination on the basis of differences in gender, ethnic origin and wealth.

Rationale

Variation within and among agroecosystems is enormous. Daily, seasonal and longer term changes in the spatial structure of agricultural biodiversity are apparent at the broad landscape level right down to small plots of cultivated land. These spatio-temporal dynamics have major implications for the way agricultural biodiversity is managed,-how, by whom and for what purpose.

Uncertainty, spatial variability and complex non-equilibrium and non linear ecological dynamics emphasise the need for flexible responses, mobility and local level adaptive resource management in which local users of agricultural biodiversity are central actors in analysis, planning, negotiations and action.

This calls for far greater appreciation of local farming practices and knowledge used by rural people to manage agricultural biodiversity in forests, wetlands, fields, rangelands, coastal zones and freshwater systems. It frequently suggest new practical avenues for technical support in which land users' own priorities, knowledge, perspectives, institutions, practices and indicators gain validity.

STAND MANAGEMENT TO MAINTAIN BIODIVERSITY

This chapter describes how specific objectives for maintaining stand structure, tree and vegetation species composition, and coarse woody debris can be determined. The recommendations in this chapter should be applied in the preparation of:

- forest development plans
- silviculture prescriptions and stand management prescriptions
- logging plans, fire management plans, and range use plans.

Objectives for wildlife trees and coarse woody debris must be included in the contents of a Forest Development Plan (OPR 15 (7) (b)). A Forest Development Plan (OPR 15 (2)) and a silviculture prescription (OPR 39 (2) (w)) must contain a reasonable assessment of non-timber resource values known to be on or adjacent to the plan area and must describe the actions to be taken to accommodate those values.

Stand level practices in this chapter are the recommended minimum requirements needed to meet the structural characteristics of natural openings.

When wildlife tree patches (group reserves) are larger than 2 ha (a patch that is isolated within the cutblock boundary and not included within the net area to be restocked) and also meet the age and structural requirements of old seral forest then these larger, within-block patches can contribute to old-seral stage forest requirements within the landscape unit and be used in landscape level retention calculations). A wildlife tree patch is synonymous with a group reserve in silvicultural terminology.

Given the high degree of ecological variability in our forests, managers need to consider biological diversity on a site-specific basis in order to most effectively apply the recommendations presented in this chapter. This variation may require exceeding the minimum retention requirements recommended in this guidebook.

The recommendations for maintaining biodiversity are an important component of ecosystem health and will have to be integrated with other objectives for forest health.

MAINTAINING STAND STRUCTURE

Stand level recommendations are designed to maintain or restore, in managed stands, important structural attributes such as wildlife trees (including standing dead and dying trees), coarse woody debris, tree species diversity, and understorey vegetation diversity.

SAFE WORK PRACTICES

Safe work practices, as established in conjunction with Workers' Compensation Board, must be followed at all times when implementing the recommendations in this guidebook. Forest workers should have freedom to remove obstacles whenever necessary to maintain a safe working environment. Trees that are marked to leave or are outside the cutblock boundaries can be felled for safety reasons, but should be left on the ground as future coarse woody debris. This will retain the benefit these trees can have on site and remove any incentives for felling and harvesting.

Wildlife trees

A wildlife tree is any standing live or dead tree with special characteristics that provide valuable habitat for conservation or enhancement of wildlife. These trees have characteristics such as large size (diameter and height) for site, condition, age, and decay stage; evidence of use; valuable species types; and relative scarcity. They serve as critical habitat (for denning, shelter, roosting, and foraging) for a wide variety of organisms such as vertebrates, insects, mosses, and lichens.

Maintaining wildlife trees within harvest and silviculture units can be ecologically beneficial in a number of ways. While standing, they provide habitat for many species (birds, bats, and other small mammals) that perform roles in maintaining ecosystem functions. Standing green trees can provide for future wildlife tree recruits. Wildlife trees will, over time, become sources of coarse woody debris and finally, through decay and nutrient cycling, become incorporated into second-growth forests.

Wildlife tree management strategies can range from the retention of existing wildlife trees, as scattered individuals or in patches, to the creation of new wildlife trees. Many approaches can be applied within a single cutblock, though retention of patches is recommended as the priority approach in most cases. Wildlife tree requirements apply to the use of all silvicultural systems.

Wildlife trees patches

Wildlife tree patches (WTP) provide several advantages over other retention strategies. Snags or other potentially dangerous trees are more easily retained in patches than as individual trees and operational inconvenience is minimized. There is also evidence that clumps of trees provide better habitat for birds than do scattered individual trees, as well as an area of relatively undisturbed forest floor within cutblocks. However where scattered individual wildlife trees already exist they should be retained.

Wildlife tree patches should be well distributed across the landscape. The maximum distance between WTPs (500 m) has been based on territory size and dispersal requirements of wildlife.

Recommendations:

Area and distribution of patches or individual trees:

- The retention of wildlife trees should be based on the pre-activity assessment of the wildlife tree values and requirements on or adjacent to the proposed cutblock; and on the described actions that are required to accommodate these values.
- The amount of retention is based on the proportion of each biogeoclimatic subzone in the landscape unit that is operable, and the degree of development that has occurred before the recommendations presented here are applied. The higher the proportion of operable area in a landscape or the more that previous development has reduced wildlife tree abundance, the greater the amount of retention required.
- Where landscape unit objectives have not been established, the application should be based on an interim landscape unit or a portion of a forest development plan that forms a contiguous geographic unit.
- Suitable areas outside the cutblock, such as riparian management areas, can contribute to the required retention provided the inter-patch spacing requirements are met, and they are mapped and designated as wildlife tree patches in the forest development plan. It is assumed that up to 75% of wildlife tree patch area requirements on the coast, and up to 50% in the interior, will be met in riparian management areas and other constrained areas.
- Good candidates for retention: individual live wildlife trees that provide special wildlife habitat value (such as nest trees); veterans; and other large trees. Patches of deciduous or unmerchantable trees provide opportunities to accomplish this objective. Defective trees of the largest size are often most valuable. These large, live, unhealthy trees (Class 2 trees that are well branched) are important as future habitat since they provide recruitment wildlife trees over time.

- Wildlife tree patches can range in size from individual trees to patches of several hectares. Individual trees, outside patches, can contribute to the required retention area on a basal area equivalency basis. However, simple stem basal area does not equate to patch area. For example, a 1 ha patch containing trees, shrubs, rocky or wet sites, and other ground level diversity is not ecologically equivalent to "x" number of single green stems (such as seed trees) distributed around a harvest block. Therefore, the use of individual trees to contribute to required retention area targets must be planned and selected carefully on a site-specific basis.
- Single green trees chosen for retention should exhibit some characteristics of a valuable recruitment wildlife tree (such as being large and well branched). In order to optimize their habitat value within the harvest block, these trees should be located, wherever possible, near areas that already contain some structural elements of stand level diversity (for example, near a riparian area, rocky outcrop, gully, hardwood patch, or meadow opening). Single live trees left near a shallow gully running across a cutblock are more valuable ecologically than those same green trees distributed randomly throughout the block.
- Wildlife tree patches should be mapped and recorded as part of the documentation of the silvicultural or stand management prescription for the cutblock and be removed from the net area to be restocked.
- No timber harvesting is permitted in designated wildlife tree patches.
- The importance of wildlife tree patches within cutblocks increases with the size of the cutblock. Patches should generally be centered around the most suitable trees and distributed throughout the cutblock, with distances between patches (or other suitable leave areas outside the block) not to exceed 500 m.
- Patches should be located in a way that minimizes windthrow risk. Advantage can also be made of locating patches along small streams, wet areas, gullies or other locations that are likely to pose harvesting, yarding or regeneration problems.
- When partial cutting silvicultural systems are used, sufficient leave trees should be retained throughout the rotation, to meet retention objectives.
- Wildlife trees must be retained at least until other suitable trees can offer equivalent replacement habitat and structural diversity. In most cases this will take at least one rotation.

Patch and live tree retention characteristics:

- A range of tree diameters should be included within patches, favouring larger stems when possible. Care should be taken to include the upper 10% of the diameter distribution of the stand (taken from cruise data), because these are the most valuable wildlife trees.
- Both live and dead trees (subject to safety requirements) should be included in patches representing a range of wildlife tree classes.
- A variety of tree species, including deciduous, should be represented.
- When possible, trees showing wildlife use or presence of heart rot, and those with a large size and well-branched structure should be retained.

MANAGEMENT PRINCIPLES FOR WILDLIFE TREES

- Wildlife tree management includes both the retention of suitable wildlife trees at the time of harvest and during silvicultural activities, and the provision for recruitment of suitable replacement wildlife trees over the rotation period.
- Generally, the most operationally feasible and biologically advantageous method for retaining wildlife trees is to leave patches of live and dead trees as no work zones and to exclude these from the net area to be reforested.
- The amount of wildlife tree area required for a specific cutblock depends on the level and distribution of existing and planned harvesting in the surrounding landscape.

This is a one-time calculation for each biogeoclimatic subzone within the landscape unit (or interim landscape unit or portion of a forest development plan forming a contiguous geographic unit) unless the landscape unit objectives change, a new landscape unit is designated, or operability limits change (changing the area available for harvest). A separate prescription is made for each subzone within the landscape. X-axis numbers (columns) are the proportion of the subzone within the landscape unit (or interim landscape unit or portion of a forest development plan forming a contiguous geographic unit) that is identified as available for harvest (that is, not inoperable or in some sort of reserve status, such as riparian or protected area).

Y-axis numbers (rows) are the proportion of the available landscape (above) that has already been harvested without application of this guidebook's recommendations or similar prescriptions.

Example: For each biogeoclimatic subzone in the landscape unit, calculate the area available for harvest (the X-axis). For example, if 30% of the SBSmc area is available for harvest, then, using the 30% column, the recommended minimum proportion of each cutblock to be managed for wildlife trees is between 1 and 9%. If 50% of the available area has already been harvested without application of these or similar guidelines (Y-axis), then 5% of each new cutblock would need to be left for wildlife tree patches. This can be distributed adjacent to cutblocks in riparian or other long-term leave areas when feasible.

Creating wildlife trees

One method of creating wildlife trees is to high-cut stumps during the use of feller bunchers. This creates standing dead trees called "stubs". These provide structure within second-growth forests and create future coarse woody debris. They cannot, however, replace all attributes associated with full height wildlife trees and thus can not be used as a complete substitute.

Recommendations:

- If mechanical harvesters are being used, snags and cull trees that must be felled should be left as high as can be reached safely. For dead trees, the maximum allowable height must be according to Workers' Compensation Board regulations.
- Stems suitable for stub creation should have some visible defect (such as canker, scar or conk) in the lower bole and little or no lean.
- Creation of wildlife trees by topping, blasting, and other methods should also be considered.

Other stand structure recommendations

Any action that encourages structural diversity within managed stands can have value to biodiversity.

Recommendations:

- Some vertical and structural variability should be maintained throughout stands or rangelands. Within larger stands, this can be achieved by the use of variable stocking densities at planting, in combination with spacing or thinning activities. In rangelands, components of shrub communities should be retained.
- In vegetation management or thinning treatments, patches or strips can be left unthinned on some sites so that dense patches develop.
- Wider spacing of patches can be used to maintain a partially open canopy that will promote understorey vegetation.
- Non-merchantable defect trees should be left as recruitment snags rather than removed.

- Patches of advance regeneration can provide structural diversity.
- Range management should be designed to maintain a component of shrubs within rangelands, based on the shrub component found in undisturbed rangelands.
- Forest encroachment on rangelands can be reduced through management practices such as prescribed fires.
- During tree pest and disease treatments, some areas should be left untreated.
- Openings should be irregularly shaped to most closely reflect natural disturbance patterns.

MAINTAINING TREE AND VEGETATION SPECIES COMPOSITION

The maintenance of the diversity of naturally occurring plant species is key to the maintenance of biological diversity within landscape units. Within cutblocks, several actions can help maintain that diversity.

Recommendations:

- The variety of native understorey plants and plant communities should be maintained across the landscape units.
- Vegetation management treatments can be designed to create variability among or within treatment areas. Effects on non-target plants should be minimized.
- A component of the deciduous species, both immature and mature, should remain after harvesting, spacing, vegetation management, and site preparation activities.
- Extensive conversion from climax to young seral species (such as Douglas-fir to lodgepole pine) or from young seral to climax should be avoided.
- Where suited to the site, stands should be regenerated with a mixture of tree species (natural and planted) rather than with a single species.
- Where mature hardwoods form a minor component of the stand (<20%), greater emphasis should be placed on maintaining these either singly or in clumps.
- Minor tree species such as yew, birch, alder, aspen and cottonwood should be maintained.

MAINTAINING COARSE WOODY DEBRIS

Maintaining coarse woody debris after harvesting is a critical element of managing for biodiversity. However, it is recognized that this requirement conflicts with existing utilization standards. Work is under way to resolve this policy conflict. Until this issue is resolved, utilization standards will take precedence over requirements for coarse woody debris.

Despite this policy conflict, some existing practices can be modified to help address the requirements for coarse woody debris. For example, preliminary indications are that post-logging residue and waste can meet the volume requirements for coarse woody debris if it is well distributed across the entire stand. This will not be the case in situations of whole-tree harvesting, clean site preparation practices, or excessive salvage of material not considered merchantable under current utilization standards.

Recommendations

- Modify whole-tree harvesting by limbing and topping on site.
- Maintain residue and waste well-distributed across the stand (avoid practices such as piling and burning).
- Leave non-merchantable material on site.

BENEFICIAL AND EFFECTIVE MICROORGANISMS

Agriculture in a broad sense, is not an enterprise which leaves everything to nature without intervention. Rather it is a human activity in which the farmer attempts to integrate certain agroecological factors and production inputs for optimum crop and livestock production. Thus, it is reasonable to assume that farmers should be interested in ways and means of controlling beneficial soil microorganisms as an important component of the agricultural environment.

Nevertheless, this idea has often been rejected by naturalists and proponents of nature farming and organic agriculture. They argue that beneficial soil microorganisms will increase naturally when organic amendments are applied to soils as carbon, energy and nutrient sources. This indeed may be true where an abundance of organic materials are readily available for recycling which often occurs in small-scale farming. However, in most cases, soil microorganisms, beneficial or harmful, have often been controlled advantageously when crops in various agroecological zones are grown and cultivated in proper sequence (i.e., crop rotations) and without the use of pesticides. This would explain why scientists have long been interested in the use of beneficial microorganisms as soil and plant inoculants to shift the microbiological equilibrium in a way that enhances soil quality and the yield and quality of crops.

Most would agree that a basic rule of agriculture is to ensure that specific crops are grown according to their agroclimatic and agroecological requirements.

However, in many cases the agricultural economy is based on market forces that demand a stable supply of food, and thus, it becomes necessary to use farmland to its full productive potential throughout the year.

The purpose of crop breeding is to improve crop production, crop protection, and crop quality. Improved crop cultivars along with improved cultural and management practices have made it possible to grow a wide variety of agricultural and horticultural crops in areas where it once would not have been culturally or economically feasible.

The cultivation of these crops in such diverse environments has contributed significantly to a stable food supply in many countries. However, it is somewhat ironic that new crop cultures are almost never selected with consideration of their nutritional quality or bioavailability after ingestion.

As will be discussed later, crop growth and development are closely related to the nature of the soil microflora, especially those in close proximity to plant roots, i.e., the rhizosphere. Thus, it will be difficult to overcome the limitations of conventional agricultural technologies without controlling soil microorganisms.

This particular tenet is further reinforced because the evolution of most forms of life on earth and their environments are sustained by microorganisms. Most biological activities are influenced by the state of these invisible, minuscule units of life. Therefore, to significantly increase food production, it is essential to develop crop cultivars with improved genetic capabilities (i.e., greater yield potential, disease resistance, and nutritional quality) and with a higher level of environmental competitiveness, particularly under stress conditions (i.e., low rainfall, high temperatures, nutrient deficiencies, and agressive weed growth).

To enhance the concept of controlling and utilizing beneficial microorganisms for crop production and protection, one must harmoniously integrate the essential components for plant growth and yield including light (intensity, photoperiodicity and quality), carbon dioxide, water, nutrients (organic-inorganic) soil type, and the soil microflora. Because of these vital interrelationships, it is possible to envision a new technology and a more energy-efficient system of biological production.

Low agricultural production efficiency is closely related to a poor coordination of energy conversion which, in turn, is influenced by crop physiological factors, the environment, and other biological factors including soil microorganisms.

The soil and rhizosphere microflora can accelerate the growth of plants and enhance their resistance to disease and harmful insects by producing bioactive substances. These microorganisms maintain the growth environment of plants, and may have secondary effects on crop quality. A wide range of results are possible depending on their predominance and activities at any one time.

Nevertheless, there is a growing consensus that it is possible to attain maximum economic crop yields of high quality, at higher net returns, without the application of chemical fertilizers and pesticides. Until recently, this was not thought to be a very likely possibility using conventional agricultural methods.

However, it is important to recognize that the best soil and crop management practices to achieve a more sustainable agriculture will also enhance the growth, numbers and activities of beneficial soil microorganisms that, in turn, can improve the growth, yield and quality of crops.

RELATIONSHIPS TO OTHER GUIDEBOOKS

This guidebook describes future desired conditions for forests and grasslands at the landscape and stand levels. Other guidebooks that also provide direction on maintaining biological diversity at the landscape level are the *Riparian Management Area Guidebook,* the *Managing Identified Wildlife Guidebook,* and the *Regional Lakeshore Guidebook.* Riparian management areas (RMAs) and wildlife habitat areas (WHAs) can contribute to meeting the old-growth and connectivity objectives within a landscape unit. It is also likely that RMAs and WHAs will be the main building blocks for the design of Forest Ecosystem Networks.

As well as these, several other guidebooks—notably those that address terrain stability, watershed assessment procedures, community watershed management, and visual landscape management—involve land zoning and recommend certain constraints to harvesting and grazing activities.

Collectively these guidebooks provide planners with further direction on how land might be zoned into the long-term leave areas needed to minimize the effects on biodiversity of habitat fragmentation and old-growth conversion. The administrative process for establishing, varying or canceling landscape units and landscape unit objectives is described in the *Higher Level Plans Guidebook.*

SUPPORT LOCAL PARTICIPATION IN PLANNING, MANAGEMENT AND EVALUATION

Actions

- Ensure public participation of women and men (particularly farmer, herder, fisherfolk and forest dweller involvement) in the development of land use and agricultural policies and in the generation of technologies

- Ensure inclusive equitable representation (gender, class, ethnic origin, age) in the participatory activities and process
- Provide capacity building for technical and scientific personnel to foster those participatory skills, attitudes and behaviour needed to learn from farmers and rural people (mutual listening, respect, gender sensitivity as well as methods for participatory learning and action)
- Provide institutional space and incentives for professionals to understand social and cultural complexity as well as agro-ecological diversity
- Support joint problem-solving among local people, scientists and/or extension workers and the development of negotiated participatory research agendas and resource management agreements, using local criteria and indicators as well as those of outside professionals and their organisations
- Support the participatory monitoring and evaluation of national policies, land use plans, food and fibre production technologies by building on the perspectives of all social actors. Encourage the use of gender desegregated and socially differentiated local indicators and criteria in monitoring and evaluation as well as in guiding subsequent technical support, policy changes and allocation of scarce resources for agricultural biodiversity management.

Rationale

From the outset, the definition of *what* agricultural biodiversity is to be conserved, *how* it should be managed and *for whom* should be based on interactive dialogue to understand how local livelihoods are constructed and people's own definitions of well being.

Most professionals have tended to project their own categories and priorities onto local people and landscape management. In particular, their views of the realities of the poor, and what should be done, have generally been constructed from a distance. Household livelihood strategies often involve different members in diverse activities and sources of support at different times of the year. Many of these, like collecting wild foods and medicine, home gardening, common property resources, share-rearing livestock and stinting are largely unseen by outside professionals.

Agricultural Research and Development and land use planning should start with enabling local people, especially the poor, to conduct their own analysis and define their own priorities. This methodological orientation is absolutely essential to meet the goals of equity, sustainability, productivity and accountability. In that context, dialogue, negotiation, bargaining, conflict resolution and joint management agreements are all integral parts of a long term participatory process which continues well after the initial appraisal and planning phases into monitoring and evaluation. This implies the adoption of a learning process approach in the management of agricultural biodiversity and its functions. It also calls for a new professionalism with new concepts, values, participatory methodologies and behaviour.

Environmental Conservation and Ecology

INTRODUCTION

Environmental conservation is an integral part of the socio economic development. The growing population, high degree of mechanisation and steep rise in energy use has led to activities that directly or indirectly affect the sustainability of the environment. The various kinds of pollution/pollutants may be broadly categorized as (a) Air pollution, (b) Water pollution, (c) Solid waste and (d) Industrial and hazardous waste. The sustainable use of bio-diversity is fundamental to ecologically sustainable development. India is one of the 12 mega diversity countries of the world. The survival and well being of any nation depends on sustainable social and economic progress that satisfies the needs and aspiration of the present without compromising the interest of future generations. Environmental problems are largely the by-products of affluence marked by resource wasteful life style. Recycling of solid and liquid wastes, bio-composting, tree planting, etc. are important for environmental conservation.

ENVIRONMENTAL COLLABORATION AND DEVELOPMENT

Collaborative or cooperative approaches to environmental and natural resource management provide potential solutions to the dilemma of the environment development tradeoff. These approaches rely on positive incentives and partnership arrangements. Over the past decade the term social capital has received considerable attention from scholars in a variety of fields. Social capital is valuable because it provides resources to solve problems of coordination and cooperation, reduces transaction costs, and facilitates the flow of information between and among individuals in community or organization. Similarly, Putnam (1993) argues that social capital makes collective works easier and, ultimately, facilitates economic and community development. The concept of social capital has become increasingly popular in a wide range of social science disciplines, but there is a lack of consensus on the meaning of term. In social science research "social capital" is used in vastly different ways. Critics have characterized research examining the impacts of social capital as '"casual empiricism", because it lacks of an obvious link between theory and measurement .

In order to better understand how social capital can help state and local governments reconcile environmental and development goals, we systematically define and classify social capital based on its scope and form. This allows us to identify different "types" of social capital that can shape collaboration and partnership among actors concerned with environment and economic development.

The Forms of Social Capital

The Uphoff (2000) suggested two dimension of social capital—structural and cognitive. Structural forms of social capital concern the roles, rules, procedures, and networks that facilitate information sharing, and collective action and decisionmaking through established roles, social networks and other social structures supplemented by rules, procedures, and precedents. As such, it is a relatively objective and externally observable construct.

Cognitive social capital refers to shared norms, values, trust, attitudes, and beliefs. It is therefore a more subjective and intangible concept (Uph off, 2000). Landry, Amara, and Lamari also classify two form of social capital: Structural and Cognitive.

They measure three type of structural social capital: Network capital, Relationship capital, and Participation capital. Cognitive social capital was measured by trust capital. Krishna (2000) makes a similar distinction between institutional capital and relational capital.

The structural (Institutional) dimension of social capital includes rule of law, formal institutions and organization structures, but it also encompasses the overall pattern of relationships in an organization and its included network. This conceptualization is similar to Granovetter's (1973) notion of weak ties.

The relational dimension of social capital concerns the nature of connections between individuals. It is characterized by levels of trust, shared norms and perceived obligation, and sense of mutual identification. This conceptualization of relational social capital is similar to Granovetter's (1973) notion of strong ties. Likewise, Feiock and Tao (2002) distinguish endogenous and exogenous social capital, and examine their effects on the regional economic development partnership as one form of collective action.

The Scope of Social Capital

The scope of social capital ranges from the micro to the macro level. Analysis of social capital at the micro level is usually associated with face-to face interaction between and among individuals , and those features of horizontal relationship, such as networks of indivi duals or households, and the associated norms and trust, that generat e externalities for the community as a whole. James C oleman (1990) includes vertical as well as horizontal associations and behavior within and among organizations by expanding the unit of ob servation and introducing a vertical component to social capital.

A macro-view of social capital includes the social and political environment that shapes social structure and enables norms to develop. This view includes the most formalized institutional relationships and structures, such as the rule of law, the political regime, the court system, and civil and political liberties. This focus on institutions draws on the work of Mancur Olson (1982) and Douglas North (1990), who have argued that such institutions have a significant effect on the pattern and rate of economic development.

The phenomena related with the micro and macro level conceptualizations are complementary and their coexistence maximizes the waves of social capital on economic and social outcomes. For example, macro institutions can provide an enabling environment in which local associations can develop and flourish; local associations can sustain regional and national institutions and add a measure of stability to them.

"ENVIRONMENTAL" IN THE DISASTER CONTEXT

The environment is often seen as the agent/cause of a disaster or perhaps as the carrier. In an earthquake or a flood, for example, the "environment" behaves in ways that bring harm to the communities affected by them—one suddenly finds the environment sitting in one's living room. However, people make choices—farming practices, use and procurement of fuels, selection of building materials and sites, etc.—that significantly affect their vulnerability to environmental disasters. This view mirrors the idea that disaster is a social construct formed by the interaction of human development with natural processes. An earthquake is a disaster only when it impacts the human infrastructure. But the environment also interacts with human society in complex ways. Floods may damage natural habitats and ecosystems; forest fires may harm forest ecosystems and damage the biotic stock in an area. Yet, floods are necessary to renew and enrich riparian corridors and wetlands and to recharge aquifers; forest fires thin out undergrowth that could fuel larger fires, and they can re-vitalize biodiversity (Sauri 2004). Floods can clog wastewater treatment plants, causing the release of untreated sewage into water bodies; floods can also mobilize contaminants and industrial chemicals that then flow downstream and possibly into those same aquifers.

Thus, an "environmental" hazard may be difficult to define, and there can be a fine distinction between an environmental hazard (*i.e.*, water out of control—a flood) and an environmental resource (*i.e.*, water in control—a reservoir). It can often be a matter of perception regarding deviations about the norm—too much rain is a flood; too little is a drought. Some definitions of environmental hazard emphasize the acute and short-term at the expense of the chronic and long-term (droughts desertification, erosion), for example:

- ...extreme geophysical events, biological processes and major technological accidents, characterized by concentrated releases of energy or materials, which pose a largely unexpected threat to human life and can cause significant damage to goods and the environment.

There is a growing understanding of environmental degradation as a contributing factor in disaster effects—*i.e.*, an exacerbating factor in damage, it worsens impact on victims and makes recovery more difficult. One example: Although the largest danger facing Turkish urban areas is earthquake, numerous other hazards exist. Improper handling of solid wastes causes explosive methane build-up, endangers the physical environment, reduces property values and destroys the scenic and tourist values of highly visited areas.... Near the larger cities, many bodies of water are so polluted that they are no longer suitable for recreational use. High levels of heavy metals are found in harbour catches, and massive fish kills are common. Marine accidents release massive, toxic discharges, sometimes causing explosions that destroy buildings and facilities. Dangerous chemicals enter the urban food chain...urban rivers are polluted...agricultural chemicals and waste water have contaminated precious aquifers.

The most recent example occurred in the South Asian tsunami—long-term damage to coral reefs and degradation of mangrove swamps in some areas reduced the capacity of natural systems to absorb or cushion the kinetic energy of the tsunami surge. Deleterious effects of degraded environmental conditions are felt most keenly (though not exclusively) by the poor, residents of shantytowns, "favelas," and other marginal or hazardous areas. They are clustered on steep slopes subject to flash floods and erosion, in dwellings built of substandard materials, with poor water and waste disposal systems. Natural disaster effects can be greatly magnified by the poor environment in which these people live.

According to Pelling (2003b), there is a tendency to focus on technical and engineering issues in addressing environmental problems or issues and to discount the influence of social characteristics on susceptibility to environmental risk. This bias towards technological and physical solutions (*e.g.*, flood walls, or leachate mitigation systems) can encourage development in hazard areas when, in fact, hazards can surpass the margin of safety provided by technological solutions.

"DISASTER" IN THE ENVIRONMENTAL CONTEXT

The field of emergency management tends to focus more on harm to the human environment and the built environment and to pay less attention to the larger environment in which humans and structures exist. Also, the emphasis is on the more acute disasters (like earthquakes or chemical spills) and less on the slow-developing problems with chronic effects (*e.g.*, Minamata or acid rain) or on acute events with long-lasting consequences (*e.g.*, Bhopal, or the Tisza River). This no doubt reflects the understandable orientation of emergency management professionals to the needs of planning for and response to the immediate effects of a disaster and the desire for speedy restoration to something approaching the status quo ante. Environmental professionals take a somewhat more comprehensive view, considering not only the human and built environments but also the matrix in which they exist. Environmental concerns include not only humans but also plants and animals, water and air quality, the fate and transport of environmental contaminants, the toxicology of human and animal effects, and the exposure and vulnerability (both acute and chronic) of the affected biota.

All of these concerns can—and should—contribute in some way to the practice of emergency management before, during, and after disasters. Environmental management confronts all of these hazards, in one manner or another, and brings the full range of scientific, technical, and managerial skills and techniques to bear on preventing mitigating, or responding to their effects.

Of course, the definitions of "emergency" and "disaster" are a bit different in the environmental field: "An environmental emergency is a tanker truck full of acid overturned and spilling in the middle of town.

An environmental disaster is that same tanker spilling into a wetland or a river." Environmental hazards are not independent of other types of hazards, and one may lead to the other or make the other worse. For example, floods can degrade water quality, release chemicals and other contaminants from impoundments or containers (or even float off the containers themselves to lodge in someone else's backyard). Earthquakes can cause transportation spills, industrial chemical releases through infrastructure damage, or damage to containment.

Destruction of the World Trade Centre released asbestos, respiratory irritants, polycyclic aromatic hydrocarbons (possible carcinogens), pulverized metals, and god-knows-what-else into the atmosphere, affecting rescue and recovery workers and undoubtedly contaminating the surrounding area (Mattei). As we have seen in the example from Turkey, above, environmental hazards may only be waiting for a triggering event to make a natural disaster even worse.

The Regulatory Imperative

Starting with the National Environmental Policy Act, 28 major environmental protection laws were enacted between 1969 and 1986.

Environmental legislation since 1986 has generally focused on expanding or extending (and, in some cases, clarifying) existing laws. A number of these laws—and their implementing regulations—specifically address emergency planning and/or response in some way.

These include:

- Resource Conservation and Recovery Act (RCRA, 1976)—Requires hazardous waste facilities to prepare and maintain emergency plans to prevent or respond to releases of hazardous wastes
- Comprehensive Environmental Response Compensation and Liability Act (CERCLA or "Superfund")—includes requirements for emergency plans during cleanup actions on uncontrolled waste sites
- Clean Water Act & Oil Pollution Act (1990)—requires Spill Prevention, Control, and Countermeasures plans be prepared by certain facilities storing petroleum fuels
- Emergency Planning & Community Right-to-Know Act (EPCRA or SARA Title III)—directed states and local governments to establish planning and coordination bodies to carry out emergency planning for chemical emergencies in their jurisdictions
- Clean Air Act, section 112r, Risk Management Programme—requires certain industrial facilities to prepare an "off-site consequence analysis" for releases of certain chemicals and to prepare emergency plans in coordination with local response agencies
- Executive Order 12856 (1993)—directs Federal facilities to comply with EPCRA regarding public notification of chemical use and emergency planning

States have enacted their own set of environmental laws and regulations that parallel, enhance, and extend Federal regulations. In many cases (California and New Jersey are good examples), state regulations are more strict than Federal requirements. An example of European regulatory action, affecting more than one country, is the Seveso Directive. A 1976 explosion at a chemical plant in a small town near Milan, Italy released a large cloud of dioxin that affected a large portion of Lombardy region. The explosion and aftermath, including the botched response, led to creation of the European Community's Seveso Directive in 1982.

A central part of the Directive is a requirement for public information about major industrial hazards and appropriate safety measures in the event of an accident. It is based on recognition that industrial workers and the general public need to know about hazards that threaten them and about safety procedures. This is the first time that the principle of "need to know" has been enshrined in European Community legislation. Much of the Seveso Directive is analogous to EPCRA with additional elements addressing what are called, in the US, "worker Right-To-Know" laws. All of this happened several years prior to the Bhopal disaster, which was part of the impetus for passage of EPCRA in the United States.

Though not regulatory in nature, there are a number of international standards regarding environmental management that specify a requirement for emergency plans. The most widely recognized is the ISO 14001 standard for Environmental Management Systems, one element of which states, "The organization shall establish and maintain procedures to identify potential for and respond to accidents and emergency situations, and for preventing and mitigating the environmental impacts that may be associated with them".

The objective of the Guiding Principles for Chemical Accident Prevention, Preparedness and Response (2003), published by the European Organisation for Economic Co-Operation and Development is to "provide guidance, applicable worldwide ... to prevent accidents involving hazardous substances and to mitigate the adverse effects of accidents that do nevertheless occur." This set of principles covers much the same ground as the EPCRA, the Risk Management programme regulations, hazardous materials transportation regulations, and various environmental and safety standards in the US.

The nexus of environmental management, development, and disaster risk

Considerable research and analysis has been done by the European Union and the United Nations to illuminate the connections among environmental hazards, sustainable development strategies (especially in the poorer countries), and disaster response and management. Living with Risk (2004), produced by the UN International Strategy for Disaster Reduction, puts it most succinctly:

The environment and disasters are inherently linked. Environmental degradation affects natural processes, alters humanity's resource base and increases vulnerability. It exacerbates the impact of natural hazards, lessens overall resilience and challenges traditional coping strategies.

Furthermore, effective and economical solutions to reduce risk can be overlooked.... Although the links between disaster reduction and environmental management are recognized, little research and policy work has been undertaken on the subject. The concept of using environmental tools for disaster reduction has not yet been widely applied by practitioners. The UN International Strategy for Disaster Reduction also focuses on the transboundary nature of disasters and the importance of a "harmonized approach" to the management of pollution of river basins, seismic hazard areas, and volcanoes. This issue is perhaps less salient in the United States, due to the extent of Federal disaster management and response. Researchers in the Swedish Embassy in Bangkok have sought to link environmental programmes with disaster risk in the context of sustainable development.

They ask:

- How can investments in environmental management and sustainable development also reduce disaster risk?
- Is there a prevention dividend that accrues from wise land use planning and development programmes?
- Can prevention dividends be measured; and, how might the ability to estimate these added values enhance policy and programme planning?

Although they find evidence for positive answers to these questions, they acknowledge that more research and analysis is necessary in order to capture the rather elusive cost/benefit parameters of disaster reduction and sustainable development.

Zones of convergence

Living with Risk (2004, p. 303) outlines ways to integrate environmental and disaster reduction strategies:

- Assessment of environmental causes of hazards occurrence and vulnerability
- Assessment of environmental actions that can reduce vulnerability
- Assessment of the environmental consequences of disaster reduction actions
- Consideration of environmental services in decision-making processes
- Partnerships and regional approaches to land use and nature conservation

- Reasonable alternatives to conflicts concerning alternative uses of resources
- Advice and information to involve actors in enhancing the quality of the environment.

Within this context, there are a number of areas where environmental management and emergency management can and should interact more positively for mutual benefit and support. Both fields would benefit from continuing and supporting the current movement in the disaster community from "reactive" disaster response to active risk management and from iterative recovery to pro-active mitigation and prevention. Parallel efforts would transition the environmental field from contaminant clean-up to risk reduction and pollution prevention, from discrete issues management to environmental management systems, and from flood control to floodplain management.

Integration of sustainability considerations into disaster mitigation and recovery can exploit the considerable overlap between environmental management and disaster management. Planners and practitioners in both fields must recognize that the overall objectives of these fields implicitly promote sustainable communities. Sustainability should be considered both prospectively (in sustainable development planning and mitigation) and retrospectively (in response and recovery).

This integration would incorporate and enhance current trends towards "holistic disaster recovery" (also "sustainable recovery") that emphasize betterment of the entire community, including environmental improvement and enhancement, through the recovery process. Living with Risk (2004) is even more direct: Disaster reduction specialists should be encouraged to anticipate environmental requirements under applicable laws and to design projects that address these requirements, coordinating closely with environmental institutions. Environmental management professionals can make considerable contributions during the mitigation and recovery phases of emergency management. They can identify possible improvements and enhancements as well as things to avoid. More importantly, after enhancements or improvements are in place, they can monitor and assess environmental performance indicators to ensure that goals are met.

Environmental assessments should be integrated into emergency planning processes, following the Environmental Impact Statement model mandated by the National Environmental Protection Act. Environmental Impact Statements should (but currently do not) specifically include disaster-hazard considerations. Rapid environmental assessments should be conducted as part of disaster damage assessment and should be an integral part of response/recovery considerations. Both environmental managers and emergency managers must be cognizant of the importance of environmental justice/ equity issues in the context of hazard and vulnerability.

Hazards of any type have a disproportionate impact on the poor and disadvantaged. A number of thorny equity issues are coming to a head in the environmental management world, among them: industrial plant and landfill siting; development in industrial or depressed areas; residential settlement on slopes or in other marginal areas; higher population density; immigrants and language differences; differential access to social services and information sources. Most of these issues have not yet been adequately addressed in emergency management planning or community dialogue.

The United States (and, to a certain extent, other nations) has become sensitized to the possibility that terrorists might attack with Weapons of Mass Destruction (nerve agents, bioweapons, "dirty bombs"). The unpleasant reality is that terrorists don't have to try that hard to create death and destruction. The ubiquitous gasoline tanker would make a handy (and easily procured) bomb; there are over 20,000 chemical plants in the US that contain enough extremely hazardous materials to require reporting under EPCRA. The existence and availability of these and other so-called "weapons of convenience" will require a much closer and more explicit cooperation between environmental professionals and emergency managers to: assess the immediate and long-term threats; to identify both mitigation and response strategies; and to manage long-term recovery and clean-up operations.

ECOLOGICAL SECURITY

Spiraling population and increasing industrialisation are posing a serious challenge to the preservation of the terrestrial and aquatic ecosystems. Environmental protection is the key to ensuring a healthy life for the people. Environmental problems are on the increase and are more pronounced in densely populated cities. Creation of awareness regarding the ecological hazards among the public is absolutely essential. Environmental conservation and abatement of pollution are critical for sustainable development.

Department of Environment

The Department of Environment was created in 1995 as the nodal Department for dealing with environmental management of the State. The Department is entrusted with the implementation of major projects like pollution abatement in the Cauvery, Vaigai and Tamiraparani rivers, pollution abatement in Chennai city waterways, National Lake Conservation Programme and all aspects of environment other than those dealt with by Tamil Nadu Pollution Control Board.

One of the main objectives of the department is to implement Environmental Awareness Programme Wide publicity is being given on World Environment Day, Ozone day and on Bhogi Day to create environmental awareness among the general public. Due to the concerted efforts, the level of air and noise pollution has been brought down to the tune of 20-25% in the last three years. To create environmental awareness among the school and college studies, 1260 eco-clubs have been formed in all the districts of State involving selected educational institutions and NGOs. It is proposed to strengthen these existing eco-clubs. The outstanding NGOs, experts and individuals are honoured with environmental awards in recognition of their excellent contribution in the field of environment.

Tamil Nadu Pollution Control Board

The Tamil Nadu Pollution Control Board enforces the provisions of the Water (Prevention and Control of Pollution) Act, 1974 as amended, the Water (Prevention and Control of Pollution) Cess Act, 1977 as amended, the Air (Prevention and Control of Pollution) Act, 1981 as amended and the relevant provisions/rules of the Environment (Protection) Act, 1986 to prevent, control and abate pollution and for protection of environment. The Board functions with its Head Office at Chennai. There are 25 District Offices and 14 Environmental Laboratories established by the Board.

Monitoring of Industries

The Board has inventorised about 28,000 industries. The Board has prescribed standards for discharge of effluent, ambient air quality and gaseous emissions from various industries and the industries have to take necessary pollution control measures to meet the standards prescribed by the Board. For effective monitoring, the Board has classified the industries into red, orange and green, based on their pollution potential.

Procedure for Issue of Consent

The Board issues consent to industries in two stages under the Water Act and the Air Act for establishment and operation of industrial units. Consent to establish is issued depending upon the suitability of the site, before the industry takes up the construction activity. Consent to operate is issued after installation of effluent treatment plant and air pollution control measures, before commissioning production. Consent is issued subject to general conditions and specific conditions.

Vehicle Emission Monitoring

The Board is carrying out the vehicle emission monitoring in Chennai, Dindigul, Palani, Udhagamandalam and Chengalpattu. In addition, private agencies have been authorised by the Transport Department in Chennai city to check the emission level of the vehicles.

The Board has upgraded and computerised all its vehicle emission monitoring stations for testing diesel driven vehicles. The Transport Corporations have also been instructed to closely monitor the emission levels of their buses.

For controlling vehicular emission, cleaner fuel like unleaded petrol, petrol with 3% benzene and low sulphur fuel (0.05%) have been introduced in Chennai Metropolitan Area. Passenger cars complying with Bharat stage-II norms alone are registered in Chennai since July 2001. 2T oil auto dispensing system have been provided in retail outlets.

The Board is also participating in a research project with an non governmental organisation and the Civil Supplies Department to study the use of gas chromatograph to detect fuel adulteration.

Action has already been taken to introduce auto liquefied petroleum gas in Chennai as it is a cleaner fuel. Steps are being taken to popularise the use of liquefied petroleum gas for autorickshaws, call taxis and other private vehicles which will help in improving air quality.

(1) Vehicle Emission Monitoring Stations provided by the Board. 8 Nos.
(2) Vehicle Emission Monitoring Stations provided by the Board in MTC Depots. 6 Nos.
(3) Emission Checking Stations provided by Private Agencies. 236 Nos.
(4) Auto Liquefied Petroleum Gas Dispensing Stations commissioned in Chennai. 12 Nos.

Noise Level Monitoring

Towards controlling noise pollution in urban areas, about 52,586 air horns were removed as of December 2004 from buses and lorries throughout the State. All the districts have been declared as air horn free districts. For noise level monitoring at the district level, sophisticated noise level meters have been provided to the District Offices of the Board.

Water Quality Monitoring

Pollution of major rivers in the State is caused by the discharge of untreated sewage from the urban local bodies and panchayats and untreated or partially treated effluent from industries. In case of industrial pollution, it is the responsibility of the industrial units to provide the required effluent treatment plants either individually or collectively so as to achieve the standards. Various pollution abatement schemes are being implemented under the National River Conservation Programme under the coordination of the Department of Environment.

Water Quality Monitoring Programmes

Under the Global Environmental Monitoring System, the Board is closely monitoring the quality of water in the Cauvery basin at Mettur, Pallipalayam, Musiri and ground water quality at Musiri. Similarly, water quality of rivers Cauvery (16 stations), Tamiraparani (7 stations), Palar (1 station) and Vaigai (1 station) and the three important lakes in Udhagamandalam, Kodaikkanal and Yercaud are being monitored under the Monitoring of Indian National Aquatic Resources System by the Board. The Board is continuously monitoring the Chennai city water ways to prevent pollution due to discharge of trade effluent from industries and sewage from local bodies and is collecting and analysing samples of river water and outfalls at regular intervals, since 1991.

MAINSTREAMING THE ENVIRONMENT

Global Ecology, International Institutions and the Crisis of Environmental Governance Five years later, at the June 1997 Special Session of the United Nations General Assembly dedicated to the review of UNCED's implementation, the climate was rather different. Optimism had given way to disappointment and, in some cases, there was real concern about the viability of the "sustainable development" model, which relies on a framework of action that does not fully address the causes of environmental destruction. Developed countries have been unable or unwilling to stick to their promise of increasing the aid to development to 0.7% of GDP, as agreed in Rio. Countries like the United States, the largest contributor to global warming, have not shown the will to take effective action that would show a real commitment to reduce their industrial emissions. On the other hand, developing countries refused to take any further steps without the guarantee that substantive financial resources would back them or that at least the commitments taken in Rio would be respected. The New York 1997 Declaration even recognized that the situation of the environment had deteriorated over the intervening five years, hoping modestly that more progress would be achieved by the next summit in 2002.

The meager positive results produced by the massive efforts in the field of international cooperation for the environment seem to indicate the contradictory character of this new, global "environmentalism." The purpose of this article is to demonstrate that, while originally being the potential source of a radical and transformative project, environmental concerns were ultimately reframed by the joint action of technocratic environmentalists, the international UN-related establishment and business and industry sectors to become compatible with global development.

Adopting an international political economy perspective, the article explores the interaction between state and markets in the construction of global environmental politics. It provides evidence that although there is a new consensus on the diagnosis of the problem - worldwide environmental degradation - very few commitments have been taken to alter the accumulation model and the patterns of production and consumption that contribute to this situation.

It suggests that the failure of the international system in ensuring a move towards sustainability, exemplified in New York, is linked to the very nature of the global bargain struck in Rio. By aiming to make "development" - in its more recent global phase, with its focus on globalized and ever expanding production, trade and consumption - become "sustainable," the concept of sustainability has been stripped of most of its meaning. The inability of the international community to deal with most global environmental issues reveals the contradictory nature of the "sustainable development" consensus and demonstrates the limits of international cooperation in the name of the environment.

Origins and Dimensions of the Ecological Project

In order to understand the meaning of the transformation of environmental concerns into a widely accepted concept, it is useful to recall the original purpose of the ecological project. The ecological movement finds its origins in a protest aimed at defending the right of individuals to regain influence over their ways of living, of producing, and of consuming. As stressed by Gorz (1992), it started as a radical cultural movement, as an attempt by individuals to control and understand the consequences of their actions. With the ecological critique, activists hoped to refocus attention on local knowledge and practices and to bridge the separation of humans from nature, a division that had been at the heart of the Enlightenment project.

In the 1970s, the ecological movement became a political movement, and there was an awareness that the demands of ecology were not only sectorial and local aspirations but rather represented a value shared across national divides (Smith 1996; Gorz 1992).

The publication of the report "Limits to Growth" by the Club of Rome in 1972 gave a scientific backing to these cultural demands and showed the risks posed by the model of industrial growth on the future of life on earth.

The report provided a holistic view of the interrelationship between population growth, food production and consumption, the industrialization process, depletion of nonrenewable resources and waste and pollution at the global level, recognizing that waste and pollution are not only a problem for the living conditions and consumption patterns of the population, but affect the very basis of the productive sphere's reproduction.

For the first time, environmental degradation provoked by economic growth was considered from a global perspective, going beyond the occasional questioning of pollution problems during the 1950s and 1960s. In addition, the report launched a real debate on the morality of growth and of the differences in consumption and living standards between developed and developing countries. The 1970s also represented an inflection in the history of social mobilization and collective action with the emergence of the "new social movements," which identify themselves as value movements carrying universal interests going beyond class, nation, sex and race borders. The new social movements such as the environmental movement appear as "modern" in the sense that they are based upon the belief that history's course can be changed by social actors and are not determined by what Touraine calls a "metasocial principle" (Offe 1988, 219). Environmentalists believe that, although representing a real challenge to our present lifestyles and habits, it is possible to move towards a sustainable society that respects nature and privileges well-being over accumulation. Speaking about the existence of a unique and unified "green movement" is clearly incorrect. Environmental concerns mean different things to different people, take many forms and are expressed through different channels. In addition, environmentalism takes very different forms in developed or in developing countries. It can mean fighting for an even better quality of life in advanced countries, and fighting for subsistence or even survival in poor countries. Despite this diversity, for the purpose of academic inquiry, three main components of the "green movement," albeit sometimes overlapping, can be distinguished. These three categories should be viewed as "ideal-typical" and not necessarily mutually exclusive.

The first tendency of the ecological movement, deep ecology, is typically a postmodern movement.

In philosophical terms, deep ecology challenges the separation between humans and nature that was at the heart of modern humanism. Deep ecology is not "anthropocentric," it is "ecocentric." As observed by Merchant (1992), it seeks a total transformation in science and in worldviews that will lead to the replacement of the mechanistic paradigm (which has dominated the past three hundred years) by an ecological framework of interconnectedness and reciprocity. The ideas of deep ecology have influenced (among others)

Greenpeace, the largest green NGO, which claims that humanist value systems must be replaced by supra-humanist values that place any vegetal or animal life in the sphere of legal and moral consideration. Greenpeace is therefore an example of an environmental organization which, based on scientific reports and examinations, acts to change worldviews and consciousness in order to promote a shift to "ecocentrism" rather than trying to act to transform the production systems which lie at the root of environmental problems.

Yet, while having influenced the most well-known environmental NGO, deep ecology remains a fairly marginal wing of the green movement. Deep ecologists have been criticized for their lack of a political critique, failing to recognize that the idea itself of "ecocentrism" is "anthropocentric." As stressed by Merchant, deep ecologists take the character of capitalist democracy for granted rather than submitting it to a critique. Their tendency to refuse to consider economic policy and to assume a purely conservationist standpoint relegates them to a secondary position.

The second component of the "green movement" is what can be called the "social ecology" movement, which is to a large extent composed of people from the "New Left," dissatisfied with Marxism. Contrary to the deep ecologists, social ecologists maintain an anthropocentric perspective: the concern for nature is understood as a concern for the environment of human beings. Social ecologists seek transformations in production and reproduction systems, that is, a transformation of political economy, as the way to achieve sustainability, social equity and well being. Social ecologists see a contradiction between the logic of capitalism and the logic of environmental protection. For them, environmental protection cannot be made dependent upon economic development, because development, in its liberal sense, has meant the subordination of every aspect of social life to the market economy, and can therefore no longer be considered as a desirable goal. The hegemonic view on "sustainable development," which rehabilitates development as the global goal of humans, is thus unsatisfactory. Social ecologists call for a rethinking of the theoretical basis of development that should include not only economic but also political and epistemological dimensions, such as the questions of participation, of empowerment and local knowledge systems.

For them, what makes development "unsustainable" at the global level is the pattern of consumption in rich countries. Thinking about sustainability thus implies considering the contradictions imposed by the structural inequalities of the global system. Finally, social ecologists vary to a certain extent in the North and in the South: generally speaking, organizations in the North sometimes carry their rejection of development as far as to strike postmodern stances, while organizations in the South focus more on equity and on the need to redistribute the benefits of development.

Finally, there is a more technocratic tendency to the green movement, a tendency that tries to make economic growth and environmental protection appear as compatible goals, which need not require a profound change in values, motivations and economic interests of social actors, nor new models of economic accumulation. For them, it is because capitalist production methods and life standards are not developed enough that environmental problems emerge.

The evidence is that environmental standards are higher in richer countries. Technocratic environmentalists seek to preserve the environment through the establishment of international institutions, the use of economic and market instruments and the development of clean and "green" technology. The result is a rather apolitical approach and activists who, though still interested in environmental protection, are not primarily committed to ideas of equity and social justice, or at least not as committed as social ecologists (Gudynas 1993). The technocratic tendency is thus essentially a rich country tendency, although it is also present in some elite circles in the South. These environmentalists tend to focus on issues of population for example, arguing that the biggest threat to the environment comes from high population growth in the Third World and the pressure it will bring to bear on the stock of natural resources. Technocratic environmentalists usually tend to belong to organizations which have little or no membership, and rely on their technical and legal expertise and on their research and publishing programs to influence decision-making. Through their close relationship with government and other influential actors and their easy access to international organizations, these organizations tend to have a greater impact than activist membership organizations.

Today, it can be said that this technocratic approach appears to be prevailing over both the biocentric (deep ecology) and the social ecology perspectives and has become what is today mainstream environmentalism, which finds its major expression in the concept of "sustainable development." Despite the challenging and radical nature of ecological concerns, the fact that they might present a potential for change in the present economic model, they were ultimately reframed so as to constitute what appears as an apolitical, techno-managerial approach.

The Formation of a Consensus on "Sustainable Development" It is interesting to examine how the apparent consensus around the concept of "sustainable development" was built and how the project of global environmental "management" became hegemonic. Two main actors have contributed to the hegemony of the liberal environmental management project.

One is the scientific and policy-making environmental community, or, in the words of Peter Haas, the environmental "epistemic community" (Haas 1990); the other actor is business and industry.

The Brundtland Report, the United Nations Conference and the Global North-South "Bargain" International environmental politics did not emerge in the 1990s. As early as 1972, a United Nations Conference on the Human Environment took place in Stockholm, launching the era of international environmental negotiations.

Stockholm did produce some significant outcomes, leading to the creation of the United Nations Environment Program (UNEP), based in Nairobi, which coordinates environmental action within the United Nations. The context of the Stockholm Conference was not very favorable to the adoption of strong environmental commitments. Developing countries were unsatisfied with the UN system and preparing the movement for a New International Economic Order.

They were not willing to yield part of their sovereignty over natural resources in the name of environmental protection, and denounced the emergence of "eco-imperialism." The oil crisis of the 1970s relegated environmental protection to a marginal position in international relations. In the 1980s, the international climate started to change as the debt crisis was seriously affecting developing countries and their role and participation in international fora. In this context, "international commissions" were established to try to elaborate global proposals to promote peace and development, such as the Brandt Commission. Efforts were also undertaken to replace environmental protection on the international political agenda. The World Commission on Environment and Development was established in 1983 under the presidency of Gro Harlem Brundtland, and asked to produce a comprehensive report on the situation of the environment at the global level.

The work of the Commission represented a landmark in international initiatives to promote environmental protection as it produced the concept of sustainable development, a concept that would become the basis of environmental politics worldwide. Sustainable development is defined by the Brundtland Report as a development that is "consistent with future as well as present needs" (World Commission on Environment and Development 1987).

The concept of sustainable development was built as a political expression of the recognition of the "finiteness" of natural resources and of its potential impact on economic activities. Indeed, the report argues that, while we have in the past been concerned about the impacts of economic growth upon the environment, we are now forced to concern ourselves with the impacts of ecological stress - degradation of soils, water regimes, atmosphere and forests-upon our economic prospects. The report offered a holistic, global vision of today's situation by arguing that the environmental crisis, the developmental crisis and the energetic crisis are all part of the same, global crisis. It offers solutions to this global crisis, which are mainly of two kinds. On the one hand there are solutions based on international cooperation, with the aim of achieving an international economic system committed to growth and the elimination of poverty in the world, able to manage common goods and to provide peace, security, development and environmental protection. On the other hand, come recommendations aiming at institutional and legal change, including measures not only at the domestic level but also at the level of international institutions. The report emphasizes the expansion and improvement of the growth-oriented industrial model of development as the way to solve the global crisis.

The Brundtland Report also promoted the view that global environmental degradation can be seen as a source of economic disruption and political tension, therefore entering the sphere of strategic considerations. For the Brundtland Commission, the traditional forms of national sovereignty are increasingly challenged by the realities of ecological and economic interdependence, especially in the case of shared ecosystems and of "global commons," those parts of the planet that fall outside national jurisdictions. Here, sustainable development can be secured only through international cooperation and agreed regimes for surveillance, development, and management on the common interest.

For example, the consequences of climate change such as rising sea levels and the effects of temperature variations on agricultural production would require deep changes in the economy and impose high costs on all countries, thus leading to very unstable situations. The issue of forest preservation can also fit into this context, since forests contribute to the stability of climate by acting as carbon sinks, and assure the regeneration of ecosystems by providing reservoirs of biological diversity. Preserving forests then becomes more than an ecological concern: it is also a security imperative. So the "environmental security" discourse was also a cause for the need to find a "consensual solution" to issues of environmental protection. The United Nations Conference on Environment and Development (UNCED), held in Rio de Janeiro in June 1992, marked the official institutionalization of environmental issues in the international political agenda.

Twenty years after the 1972 Stockholm Conference, which was on the "Human Environment," Rio meant a real shift in the vision that had dominated environmental politics so far. After Rio, environmental considerations became incorporated into development, and a "global bargain" was struck between North and South on the basis of the acceptance from both sides of the desirability of achieving a truly global economy which would guarantee growth and better environmental records to all. UNCED recognized the "global finiteness" of the world, i.e., the scarcity of natural resources available for development, but adopted the view that, if the planet is to be saved, it will be through more and better development, through environmental management and "eco-efficiency." The UNCED process involved over a hundred and fifty hours of official negotiations spread over two and a half years, including two planning meetings, four Preparatory Committees (Prepcoms), and the final negotiation session at the Rio Summit in June 1992.

The major result of UNCED is called "Agenda 21," a 700-page global plan of action which should guide countries towards sustainability through the 21st century, encompassing virtually every sector affecting environment and development. Besides Agenda 21, UNCED produced two non-binding documents, the "Rio Declaration" and the Forest Principles. In addition, the climate change and the biodiversity conventions, which were negotiated independently of the UNCED process in different fora, were opened for signature during the Rio Summit and are considered as UNCED-related agreements. The "Rio Declaration," which was the subject of much dispute between the Group of 77 (the coalition of developing countries) and industrialized countries, mainly the United States, illustrates well the kind of bargain reached in Rio.

It recognizes the "right of all nations to development" and their sovereignty over their national resources, identifies "common but differentiated responsibility" for the global environment, and emphasizes the need to eradicate poverty, all demands put forward by the Group of 77. In return, the suggestions by the G77 to include consumption patterns in developed countries as the "main cause" of environmental degradation and the call for "new and additional resources and technology transfer on preferential and concessional terms" were rejected by OECD countries.

In the end, on the issue of finance, an institution called the "Global Environment Facility" (GEF) was set, under the joint administration of the World Bank, the United Nations Development Program (UNDP) and the United Nations Environment Program (UNEP), as the only funding mechanism on global environmental issues, and OECD countries committed themselves to achieving a target of 0.7 percent of GNP going to ODA

(Overseas Development Assistance) by the year 2000, to help developing countries implement UNCED's decisions.

Despite the failure of the G77 to win significant concessions on financial resources, if one considers the differences in priorities between developed and developing countries and the conflictual character of the negotiation process, UNCED's outcomes were still seen by the international establishment as quite impressive, marking "an important new stage in the longer-term development of national and international norms and institutions needed to meet the challenge of environmentally sustainable development."

A Commission on Sustainable Development (CSD) was established to monitor and report on progress towards implementing UNCED's decisions. In particular, the CSD's stated aims are to enhance international cooperation by rationalizing the intergovernmental decision-making capacity, and to examine progress in the implementation of Agenda 21 at the national, regional and international levels.

After UNCED, environmental considerations were "integrated" at all levels of action. The "sustainable development paradigm," as some authors recognize, is already replacing the "exclusionist paradigm" (i.e., the idea of an infinite supply of natural resources) in some multilateral financial institutions, as well as in some state bureaucracies and in some parliamentary committees. Most economists now acknowledge that natural resources are scarce and have a value that should be internalized in costs and prices.

Organizations such as the European Union made the "integration" of environmental concerns one of their leading policy principles.

Many countries carried out environmental policy reform to implement UNCED's decisions and the Agenda 21. The boundaries of environmental politics were broadened and its links with all other major issues on the international arena, such as trade, investments, debt, transports, for example, were examined. Efforts were also undertaken to improve environmental records of multilateral finance and development institutions. The World Bank, which has a long history of contributing to environmental degradation by financing destructive projects, went through a "greening" process, and now has a "Department of the Environment" which conducts "environmental impact assessments" and imposes "environmental conditionalities" before granting loans. The World Trade Organization has a "Committee on Trade and Environment" (CTE) which is in charge of ensuring that open trade and environmental protection are mutually supportive. All these efforts can be seen, according to Porter and Brown (1996), as part of a longer-term process of evolution toward environmentally sound norms governing trade, finance, management of global commons, and even domestic development patterns. Environmental considerations were then to be introduced in all major international bureaucracies as a dimension to take into consideration in decision-making processes, and as a challenge for global management. To a certain extent, the "technocratic" approach became hegemonic because it best suited the interests of the international development elite as it magnified its managerial responsibilities. In a time when the legitimacy and utility of the United Nations system was being seriously questioned by its idealizer and major financial supporter - the United States - the goal of making environment and development compatible was seized by some UN agencies as an unexpected opportunity to regain credibility, as well as to be granted funds and to hire new staff for recently created units on "trade and environment" or "finance and environment." UNCED provided a new legitimacy to international organizations such as the World Bank or the World Trade Organization and to their bureaucracies, which now try to assume a leading role in "managing the earth." With the promotion of economic growth to a planetary imperative and the rehabilitation of technological progress, both development institutions and organizations and states appeared as legitimate agents to solve global environmental problems.

If international organizations have benefited from the global perspective that emerged from Rio, they have also contributed to mold it. There is an active "epistemic community," which includes both the international organization establishment and large environmental NGOs, promoting the "global environmental management" approach.

These groups tend to believe that their moral views are cosmopolitan and universal, and emphasize the existence of an international society of human beings sharing common moral bonds. In this kind of "same boat" ideology, environmental concerns tend to be presented as moral imperatives, related neither to political nor to economic advantages. It would be a consensual concern, a sort of universal principle accepted over borders and political boundaries. An example of an institution promoting these ideas is given by the Commission on Global Governance. In the words of the Commission, "we believe that a global civic ethic to guide action within the global neighborhood and leadership infused with that ethic are vital to the quality of global governance. We call for a common commitment to core values that all humanity could uphold. We further believe humanity as a whole will be best served by recognition of a set of common rights and responsibilities."

Part of the Green movement came to support this "same boat ideology" and was incorporated into the epistemic community. Actually, mainstream conservationist environmentalists were fully admitted into the global environmental management establishment, conferring legitimacy to the UNCED process.

NGOs contributed to UNCED to a degree unprecedented in the history of UN negotiations. NGOs lobbied at the official process, participated in Prepcoms and were even admitted in some countries' delegations, a novelty which was rendered possible by resolution 44/228 calling for "relevant non-governmental organizations in consultative status with the Economic and Social Council to contribute to the Conference, as appropriate."

In addition, during UNCED, NGOs organized in Rio a meeting which ran parallel to the official governmental conference. The "Global Forum," which gathered about 30,000 people, represented 760 associations, among participants and visitors, in a sort of "NGO city." During one week, the Global Forum became home to environmentalists and social activists, to Indians and ethnic minorities, and to feminists and homosexual groups, all united to "save the earth." NGOs organized many demonstrations protesting against the modest results of the official summit and elaborated their own agenda for improving environmental protection worldwide.

Yet, in the eyes of some observers, NGO efforts tended to become coopted by larger and richer groups from advanced countries, which had more means, not only financially but also in terms of organizational, scientific and research capac- ity, to promote their own views (Chatterjee and Finger 1994).

In the end, NGOs decided that they would sign, in Rio, NGOs "treaties" on all the issues being discussed at the UNCED official meeting. The main activity at the Global Forum was then the "treaty negotiation" process, just like at the official forum, a process which proved to be very disappointing, as the same North-South conflicts that were blocking UNCED tended to separate northern and southern NGOs.

Ultimately, the NGO treaty process was little more than a pantomime of real diplomacy, and ultimately, the treaties agreed upon, negotiated among a couple of dozen NGOs, had a very modest impact on the future of NGO activities.

The representation at the Global Forum was also very unequal, illustrating differences in means between northern NGOs, very present, and southern NGOs. Asian, and above all, African NGOs, were severely under-represented. Differences in associative traditions and language barriers also explain the hegemony of Anglo-Saxon organizations at the Global Forum. In the end, influential NGOs decided to concentrate their efforts on lobbying the official conference. The Earth Summit in 1992 thus represented a real moment of acceleration for NGO activities, as it allowed some of them to have a better idea of what their counterparts were doing in other parts of the world, and was the base for establishing cooperation projects and partnerships among organizations. Yet while NGO efforts illustrated by the Global Forum aimed at uniting NGOs worldwide, the green movement came out of Rio appearing even weaker and more fragmented, with the polarization between "realist," co-operative NGOs on the one side and "radical," transformative NGOs on the other.

Finally, the "sustainable development" approach also suited the interests of some governments in the Third World which are primarily committed to economic development and sought through UNCED to obtain concessions in financial and technological terms in exchange of their support for environmental management. Some Third World countries are still marked by a "developmentalist" ideology in which economic development comes before all else.

In addition, resource rich countries such as Malaysia, Indonesia, or Brazil, have traditionally had a vision of unending and expanding frontiers, in which land and natural resources are unlimited and no constraints are seen to exist on the use of resources. As a result, they were unwilling to accept the elaboration of international regimes aiming at limiting their sovereignty over the exploitation of natural resources.

The issue of sovereignty had long been a major source of tension during international environmental negotiations. As long ago as the Stockholm Conference in 1972 developing countries had pressed for the inclusion of a specific principle on the topic. Principle 21 of the Stockholm Declaration stated that "States have, in accordance with the Charter of the United Nations and the principles of international law, the sovereign right to exploit their own resources pursuant to their own environmental policies, and the responsibility to ensure that activities within their jurisdiction or control do not cause damage to the environment of other States or areas beyond the limits of national jurisdiction." The same debate arose when UNCED was convened, and in the end the sovereignty principle as in stood in the Stockholm Declaration's Principle 21 was included in the Rio Declaration.

In addition, a guarantee that economic development would continue to be the priority on the international agenda was an essential element for developing countries. The reaffirmation of the right to development, and of the sovereignty principle, ensured in Rio, were then the two elements that made agreement at UNCED possible for the Group of 77. The alliance between environment and development could then become official. As described by the vice-president of the International Institute for Environment and Development (IIED), "it has not been too difficult to push the environment lobby of the North and the development lobby of the South together. And there is now in fact a blurring of the distinction between the two, so they are coming to have a common consensus around the theme of Sustainable Development" (World Commission on Environment and Development 1987, 64). Yet to fully understand the nature of this consensus around sustainable development, one last actor needs to be introduced. The actor whose vision shaped most fundamentally the content of this consensus and the real winner of Rio, the business and industry sector, and in particular transnational corporations.

ECOLOGICAL FACTORS: DYNAMICS AND STABILITY

Ecological factors which affect dynamic change in a population or species in a given ecology or environment are usually divided into two groups: abiotic and biotic. Abiotic factors are geological, geographical, hydrological and climatological parameters. A biotope is an environmentally uniform region characterized by a particular set of abiotic ecological factors. Specific abiotic factors include:

- Water, which is at the same time an essential element to life and a milieu
- Air, which provides oxygen, nitrogen, and carbon dioxide to living species and allows the dissemination of pollen and spores
- Soil, at the same time source of nutriment and physical support
- Soil pH, salinity, nitrogen and phosphorus content, ability to retain water, and density are all influential
- Temperature, which should not exceed certain extremes, even if tolerance to heat is significant for some species
- Light, which provides energy to the ecosystem through photosynthesis
- Natural disasters can also be considered abiotic

Biocenose, or community, is a group of populations of plants, animals, micro-organisms. Each population is the result of procreations between individuals of same species and cohabitation in a given place and for a given time. When a population consists of an insufficient number of individuals, that population is threatened with extinction; the extinction of a species can approach when all biocenoses composed of individuals of the species are in decline. In small populations, consanguinity (inbreeding) can result in reduced genetic diversity that can further weaken the biocenose. Biotic ecological factors also influence biocenose viability; these factors are considered as either intraspecific and interspecific relations.

Intraspecific relations are those which are established between individuals of the same species, forming a population. They are relations of co-operation or competition, with division of the territory, and sometimes organization in hierarchical societies. An antlion lies in wait under its pit trap, built in dry dust under a building, awaiting unwary insects that fall in. Many pest insects are partly or wholly controlled by other insect predators.

Interspecific relations-interactions between different species-are numerous, and usually described according to their beneficial, detrimental or neutral effect (for example, mutualism (relation ++) or competition (relation --). The most significant relation is the relation of predation (to eat or to be eaten), which leads to the essential concepts in ecology of food chains (for example, the grass is consumed by the herbivore, itself consumed by a carnivore, itself consumed by a carnivore of larger size). A high predator to prey ratio can have a negative influence on both the predator and prey biocenoses in that low availability of food and high death rate prior to sexual maturity can decrease (or prevent the increase of) populations of each, respectively. Selective hunting of species by humans which leads to population decline is one example of a high predator to prey ratio in action. Other interspecific relations include parasitism, infectious disease and competition for limiting resources, which can occur when two species share the same ecological niche.

The existing interactions between the various living beings go along with a permanent mixing of mineral and organic substances, absorbed by organisms for their growth, their maintenance and their reproduction, to be finally rejected as waste. These permanent recyclings of the elements (in particular carbon, oxygen and nitrogen) as well as the water are called biogeochemical cycles.

They guarantee a durable stability of the biosphere (at least when unchecked human influence and extreme weather or geological phenomena are left aside).

This self-regulation, supported by negative feedback controls, ensures the perenniality of the ecosystems. It is shown by the very stable concentrations of most elements of each compartment. This is referred to as homeostasis. The ecosystem also tends to evolve to a state of ideal balance, reached after a succession of events, the climax (for example a pond can become a peat bog).

ECOLOGICAL THEORY AND ECOLOGICAL MODEL OF DISASTER MANAGEMENT

The basic premise of ecological theory is that systems are dynamic, in flux and that "everything is connected to everything else." Bronfenbrenner, in his groundbreaking treatise, argued that human relationships and interaction can best be understood within an ecological context conceptualized as various levels or layers of increasing organizational complexity. Numerous researchers have applied the ecological approach as a means to better describe and understand complex human behaviours.

Ecological models consist of a nested or layered arrangement of successive structural levels. In the proposed model, these layers represent the various organizational levels of disaster management. The proposed ecological model of disaster management also emphasizes the systemic and mutual interconnectedness of the various levels of disaster management during each phase of the disaster cycle In contrast to a "silo" orientation where disaster planning, preparedness, response and recovery efforts often occur independently at the various levels of organizational, this ecological model suggests that the various disaster management organizational levels are mutually interdependent. The ecological model assumes that each disaster management planning, preparedness, response and recovery level is nested within increasingly more complex organizational and contextual level(s). According to this conceptualization, each layer of the ecological model of disaster management, and ultimately the effectiveness of the disaster management efforts, depends upon the functional interactions among the various organizational levels. While this model assumes that every disaster management organizational level interacts with every other level, it also assumes that the mutual interactions are strongest between those most proximal nested level(s).

One inference this ecological model of disaster management is that the overall effectiveness of disaster preparedness response and recovery efforts will depend disproportionately upon the least prepared level involved. That is, the "weakest link" of any of the involved nested layers of the ecological model will disproportionately impair overall disaster management. In general, disaster management efforts will be diminished and less effective as a result of a lack of preparedness, planning, response or recovery efforts at any one or more of the organizational levels or layers of the ecological model.

The Levels of the Ecological Model of Disaster Management

The various levels of this ecological model of disaster management will initially be described in general terms and the various elements of disaster management will be described within the context of this model. Some specific exemplars of these ecological levels in terms of various hazards will also be offered. At the same time the interactions and mutual interdependencies of the various ecological levels of this disaster management model will be highlighted. Planning, preparedness, response and recovery elements as well as disaster factors affecting the response and recovery phases of this model will also be considered within the context of this ecological framework.

Individual / Family System Level of the Ecological Model

Most preparedness agencies recommend that individuals and families develop their own disaster plans and also recommend that they should compile resources ("disaster kits") to survive on their own for 3-7 days. Further, disaster planners recommend that families draw up formal escape and evacuation plans and practice drilling them. In the case of a potential influenza pandemic individuals and families will also need to practice basic hygiene, such as regular hand washing and to "cover their cough" to slow and deter the spread of this respiratory viral infection. Some planners even recommend disaster specific annexes for family plans, such as to "shelter in place" in response to a chemical or biological event. The degree to which individuals and families plan for and heed these recommendations will dramatically impact the disaster response at the organizational and community levels of the ecological model. A survey conducted in Florida in the aftermath of the active 2004 hurricane season documented that nearly half (48.7 per cent) of all Florida residents had no evacuation plan before any of the four major 2004 hurricanes. To provide another concrete example of the interaction between ecological levels, if health care providers feel their own families' safety is at risk or is compromised by a community-wide disaster they may be unwilling to report to work at their clinic or medical centre, limiting and adversely affecting the capacity/capability of their medical facility's disaster response.

Another important aspect of the individual level disaster of preparedness and response addresses the ability to provide assistance during the immediate post-disaster phase. One clear lesson learned from prior disasters is that true first responders are often the disaster survivors themselves. Individuals who possess first responder and emergency medical skills, such as first aid and cardiopulmonary resuscitation are able to provide aid to their own family members. Thus this model is useful in understanding how individual/family actions augment and may contribute to the organizational and community levels of the disaster response.

Organizational Level of The Ecological Model

The workplace, hospital, school community or other organizations are the next level of the ecological model of disaster management. At this level, the focus is on the organizational system, employees and facilities. In educational settings, organizational plans and preparedness efforts need to focus on the faculty, staff and students. For medical facilities the focus is on health care workers, staff and patients. Each workplace organization must have a disaster plan, evacuation routes and disaster supplies. Disaster training at the workplace can enhance workers' skills (*e.g.*, CPR, first aid), which can be critically important in the immediate aftermath of a disaster for their co-workers, their families and other disaster victims. Also, organizations such as corporations need to have disaster business continuity and recovery plans for economic, legal and ethical reasons. Regional/community/city recovery efforts in the aftermath of a disaster will depend, to a significant degree, on the economic resilience of the major employers. If downstream job losses related to a disaster are significant and prolonged, this will adversely impact the recovery of the disaster affected community, the community organizations as well as individuals and families residing or working in that community.

The hospital or medical centre is a special case of the organizational level in the context of the ecological model of disaster management. In any disaster with large numbers of casualties the ability of nearby medical facilities to screen, triage and treat disaster survivors, especially in the post-disaster surge, will be critically important. All Joint Commission on Accreditation of Health Care Organizations accredited hospitals are required to develop disaster plans and use an incident command system. The impact of a disaster on the surrounding community will be far greater if, as was the case with Hurricane Katrina, the disaster itself disrupts or even shuts down operations of one or more community medical facilities. Likewise, the long term loss of medical facilities and personnel in the aftermath of certain disasters, such as Hurricane Katrina, may have an enduring adverse impact on the health and recovery of the entire disaster affected community.

Community Level of the Ecological Model

The Red Cross maxim that "All disasters are local" generally refers to the initial responsibilities of the local cities, municipalities or counties in which a disaster occurs. However, consistent with the ecological model, the individual/family and local organization layers are nested within this more complex community/city/county layer. While certain disaster management efforts can probably best be approached at the community level, a concerted effort must also be made to encourage preparation and a partnership among individuals/families and community businesses and organizations. Local health departments, Red Cross agencies, and emergency management divisions affiliated with city and country governments are often the front-line, emergency responders in a community-wide disaster. Likewise, many communities have now trained and educated Community Emergency Response Team and Medical Reserve Corps volunteers. Another crucial role of local disaster response agencies is to assess the impact of a disaster in terms of casualties, damage to the community infrastructure and the local resources needed to respond to the disaster. Assessment information and status updates need to be communicated to state emergency management organizations to determine if, and what, state disaster response resources might be needed. Local agencies in disaster affected communities are also important resources and conduits for individuals, families and organizations in terms of disaster education, risk communication and planning guidance.

State Level of Ecological Model

When the resources and capabilities of a local community or jurisdiction are exceeded the state may provide resources and assets that support and sustain local community disaster response and recovery efforts. In each state emergency management organizations and state health departments also plan and prepare for major disasters. State agencies provide training, education and conduct exercises often in collaboration with local partners. Again, consistent with the ecological model, there needs to be a coordinated and collaborative disaster planning, preparedness and response effort involving the state and local disaster agencies to ensure an optimal disaster response. As an example of collaborative state and local disaster planning preparedness efforts, the State of Washington Department of Health recently convened an advisory group to consider policies and procedures needed to distribute potentially scarce antiviral medications in the event of an influenza pandemic. This Antiviral Advisory Planning Group included representatives from local public health departments and emergency management agencies as well as community medical centres.

American Indian Tribal Level of the Ecological Model

In considering tribal preparedness within the ecological model it is important to note that there are more than 550 federally recognized American Indian tribal organizations within the United States. The unique legal relationship between the Federal government and tribes has been codified in the US Constitution, treaties, statutes and case law.

Each tribe is unique in its culture, tradition and worldview. In terms of the ecological model, tribal disaster preparedness and response may be complicated by jurisdictional issues at the tribal nations, state and federal levels.

Gaps and overlaps in service jurisdiction are common on tribal nations, with different nations holding different views about the role of state and federal agencies on issues related to medical services, civil defence, and law enforcement. Because organizations providing such services are also involved in disaster response efforts, inter-organizational conflicts and lack of jurisdictional clarity can mean that citizens living on tribal nations may encounter service gaps during and after a disaster.

Due to these complexities and tribal values addressing autonomy and it is imperative that planning and preparedness efforts involving tribal nations include discussions among all the key stakeholders. In addition, non-tribal municipalities located adjacent to tribal nations need to assess their functions in relation to tribal-level services that also play a role in mitigating and/or responding to disasters.

Federal Level of the Ecological Model

When a disaster exceeds the capabilities of both local and state authorities, the governor in the affected state may request a presidential disaster declaration. In the event of a federal disaster declaration needed federal resources (Emergency Service Functions: ESF's 1-14) are made available to the disaster impacted communities.

The National Response Plan (NRP) and National Incident Management System (NIMS) are designed specifically to integrate federal, state and local community disaster response agencies into a unified command. The NRP, last updated May 25, 2006, is a detailed plan designed to coordinate the activities of federal, state, local, tribal, private sector and non-governmental organizations in disaster prevention, mitigation response and recovery. The NRP is an all hazards plan built on the template of NIMS, which provides a core set of principles and organizational processes.

Global level of the Ecological Model

For truly catastrophic events that require a global level response, such as the Indian Ocean tsunami of 2004, there are global assets available through the World Health Organization (WHO) as well as non-governmental organizations (NGO's), such as the International Red Cross.

Global assets and expertise were made available to aid the recoveries of countries whose ecosystems were affected by the 2004 Indian Ocean tsunami. For example, the United Nations Environment Programme created a task force to respond to requests for ecosystem(s) technical assistance from tsunami affected countries. Additionally, there are governmental agencies with disaster response and recovery missions, such as the Canadian Development Agency and the United States Agency International Development. At this level of disaster management, there is a need for cooperation and collaboration between these global agencies and the government(s) of the affected country or countries.

ECOLOGY AND THE POLITICS OF KNOWLEDGE

The paradigm of modern science has evolved in the last few centuries in an environment where all economic activities were aimed at maximising the productivity of man-made processes in individual sectors of the economy. This led to the development of modern technologies with highly negative externalities which remained invisible within the conceptual framework of modern science and economics. This shortcoming emanates from three basic fallacies of modern scientific knowledge:

1. It identifies development merely with sectoral growth, ignoring the underdevelopment introduced in related sectors through negative externalities and the related undermining of the productivity of the ecosystem.
2. It identifies economic value merely with exchange value of marketable resources. ignoring use values of more vital resources and ecological processes.
3. It identifies utilization merely with extraction. ignoring the productive and economic functions of conserved resources.

Development planning based on these false identifications tends to create severe ecological problems because of its inability to recognise ecosystem linkages and the ecological processes operative in the natural world. The ecological relationships between the sectors of natural resources contribute to essential ecological processes which are frequently found to be vital for human survival. Thus, the stability of ecological processes is not merely a matter of aesthetics. An incomplete understanding of the material and economic values of ecological processes leads to the destruction of the material conditions for economic development and eventually survival.

Since the availability of essential and vital resources for survival is dependent on the maintenance of essential ecological processes, economic activities which generate sectoral growth in the shortterm by destroying the essential ecological processes cannot lead to development in the long run. On the contrary, by decreasing the productivity and availability of vital resources, they initiate the process of underdevelopment.

When the natural world is viewed ecologically as a system of interrelated resources which maintain the material basis for human sustenance, economic values can no longer be perceived merely as exchange values in the market. Economic values in the ecological perspective are not always equivalent to their exchange value in the market, evaluated without any significance to their use value.

As a corollary, natural resources can have economic utility that cannot be quantified through the exchange value in the market. Such economic utility includes the maintenance of essential ecological processes that support human survival and, thus, all economic activities. The economic utilisation of resources through extraction may, under certain conditions, undermine and destroy vital ecological processes leading to heavy but hidden diseconomies. The nature of these diseconomies can be understood only through the understanding of ecological processes operating in nature.

The economics of sustenance and basic needs satisfaction is, therefore, linked with ecological perceptions of nature. The economics of sectoral growth on the other hand is related to reductionist science and resource wasteful technologies which are productive in the narrow context of sectoral and labour inputs, but may be counter-productive in the context of the overall economic base of natural resources.

The case studies in the following chapters are only representative of thousands of such cases seen everywhere. They reveal a certain pattern of contemporary economic development which can be identified thus:

1. Development has been equated only with the growth of manufacture in individual industrial sectors and with the increase in productivity of only man-made processes.
2. This sectoral growth of man-made processes has also led to ecological destruction of the natural resource base, affecting negatively other sectors of the economic system. This leads to the decay of systems productivity of all productive processes, man-made and natural.

As a result of this limitation of contemporary economics, economic development has, consequently, been taken to be synonymous with growth. The higher the rate of sectoral growth, the higher is the index of economic development. Possible ecological destruction caused by the resource intensity of sectoral growth that is guided purely by non-ecological economic considerations, has never been introduced in the processes of planning for economic development. The benefit-cost analysis of development projects has thus externalized those ecological changes and is incomplete in three important ways:

1. It deals with benefits and costs as profits and losses in financial ferms.
2. It deals with benefits and costs only in the narrow sectoral perspective and ignores costs generated by inter-sectoral linkages.
3. It deals with benefits that are largely available to more visible and economically powerful groups and ignores costs that are borne by the less visible and economically weaker groups. These costs and the associated underdevelopment are thus made invisible in modern economic analysis.

The utilisation and management of natural resources in India has so far been guided by the narrow and sectoral concept of productivity and restricted benefit-cost analysis. This narrow concept of productivity and benefit-cost analysis has blocked the conceptualization of the criteria of 'ationality of technology choice which maximizes needs satisfaction while ninimising resource use, thus maximising systems productivity.

For example, the clear felling of natural forests in the catchments of 'ivers, and planting of industrial species of trees has been justified on the ;rounds of increasing productivity of forests.

This concept of productivity is, however, only related to productivity of industrial timber, while forests produce other forms of biomass, like fodder ınd green mulch, or maintain productivity of soil and water resources. The lirect impact of the clear felling of catchment forests on agricultural production hrough its destructive impact on soil and destabilisation of the hydrological palance is not taken into account in the calculations of the benefits and costs ıssociated with forests.

Regular floods and droughts, which are the consequences of irrational and and water management, are branded as natural disasters for which the vhole nation pays heavily.

Consequently, the poor and marginal groups which depend on agriculture for their livelihood face increasing impoverishment and poverty. This thrusting of negative externalities on the poor and marginal groups directly leads to the polarisation of society into two groups. One group gains from the process of narrow sectoral growth, while the poor and marginalised majority suffer because of the ecological destruction of natural resources on which they depend for survival.

The dialectical contradiction between the role of natural resources in production processes to generate growth and profits and their role in natural processes to generate stability is made visible by movements based on the politics of ecology.

These movements reveal that the perception, knowledge and value of natural resources vary for different interest groups in society.

The politics of ecology is thus intimately linked with the politics of knowledge. For subsistence farmers and forest dwellers a forest has the basic economic function of soil and water conservation, energy and food supplies, etc. For industries the same forest has only the function of being a mine of raw materials. These conflicting uses of natural resources, based on their diverse functions, are dialectically related to conflicting perceptions and knowledge about natural resources. The knowledge of forestry developed by forest dwelling communities therefore evolves in response to the economic functions valued by them. In contrast, the knowledge of forestry developed by forest bureaucracies, which respond largely to industrial requirements, will be predominantly guided by the economic value of maximising raw material production.

The way nature is perceived is therefore related to the pattern of utilisation of resources. Modern scientific disciplines which provide the currently dominant perspectives of nature have generally been viewed es 'objective', 'neutral' and 'universally valid'. These disciplines are, however, particular responses to particular economic interests. This economic determination influences the content and structure of knowledge about natural resources which, in turn, reinforces particular forms of resource utilisation The economic and political values of resource use are thus built into the structure of natural science knowledge.

PRINCIPLE OF ECOLOGY AND ECOSYSTEM

The first principle of ecology is that each living organism has an ongoing and continual relationship with every other element that makes up its environment. An ecosystem can be defined as any situation where there is interaction between organisms and their environment.

The ecosystem is of two entities, the entirety of life, the biocoenosis, and the medium that life exists in, the biotope. Within the ecosystem, species are connected by food chains or food webs. Energy from the sun, captured by primary producers via photosynthesis, flows upward through the chain to primary consumers (herbivores), and then to secondary and tertiary consumers (carnivores and omnivores), before ultimately being lost to the system as waste heat.

In the process, matter is incorporated into living organisms, which return their nutrients to the system via decomposition, forming biogeochemical cycles such as the carbon and nitrogen cycles. The concept of an ecosystem can apply to units of variable size, such as a pond, a field, or a piece of dead wood. An ecosystem within another ecosystem is called a micro ecosystem. For example, an ecosystem can be a stone and all the life under it. A meso ecosystem could be a forest, and a macro ecosystem a whole eco region, with its drainage basin. The main questions when studying an ecosystem are:

- Whether the colonization of a barren area could be carried out
- Investigation the ecosystem's dynamics and changes
- The methods of which an ecosystem interacts at local, regional and global scale
- Whether the current state is stable
- Investigating the value of an ecosystem and the ways and means that interaction of ecological systems provides benefits to humans, especially in the provision of healthy water.

Ecosystems are often classified by reference to the biotopes concerned. The following ecosystems may be defined:

- As continental ecosystems, such as forest ecosystems, meadow ecosystems such as steppes or savannas, or agro-ecosystems
- As ecosystems of inland waters, such as lentic ecosystems such as lakes or ponds; or lotic ecosystems such as rivers
- As oceanic ecosystems.

Another classification can be done by reference to its communities, such as in the case of an human ecosystem.

THE ECONOMY OF NATURAL ECOLOGICAL PROCESSES

The terms ecology and economy are rooted in the same Greek word 'oikos' or household. Yet in the context of market-oriented development they have been rendered contradictory: 'Ecological destruction is an obvious cost for economic development'-a statement which is often repeated to ecology movements.

Natural resources are produced and reproduced through a complex network of ecological processes. Production is an integral part of this economy of natural ecological processes but the concepts of production and productivity in the context of development economics have been exclusively identified with the industrial production system for the market economy. Organic productivity in forestry or agriculture has also been viewed narrowly through the production of marketable products of the total productive process.

This has resulted in vast areas of resource productivity, like the production of humus by forests, or regeneration of water resources, natural evolution of genetic products, erosional production of soil fertility from parent rocks, remaining beyond the scope of economics.

Many of these productive processes are dependent on a number of ecological processes. These processes are not known fully even within the natural science disciplines and economists have to make tremendous efforts to internalize them. Paradoxically, through the resource ignorant intervention of economic development at its present scale, the whole natural resource system of our planet is under threat of a serious loss of productivity in the economy of natural processes.

At present ecology movements are the sole voice to stress the economic value of these natural processes. The market-oriented development process can destroy the economy of natural processes by over exploitation of resources or by the destruction of ecological processes that are not comprehended by economic development. And these impacts are not necessarily manifested within the period of the development projects.

The positive contribution of economic growth from such development may prove totally inadequate to balance the invisible or delayed negative externalities stemming from damage to the economy of natural ecological processes. In the larger context, economic growth can thus, itself become the source of underdevelopment.

The ecological destruction associated with uncontrolled exploitation of natural resources for commercial gains is a symptom of the conflict between the ways of generating material wealth in the economies of-market and the natural processes. In the words of Commoner: 'Human beings have broken out of the circle of life driven not by biological needs, but the social organisation which they have devised to 'conquer' nature: means of gaining wealth which conflict with those which govern nature."

Ecology Movements and Natural Resources

The recent period in human history contrasts with all the earlier ones in its strikingly high rate of resource utilization. Ever expanding and intensifying industrial and agricultural production has generated increasing demands on the world's total stock and flow of resources. These demands are mostly generated from the industrially advanced countries of the North and the industrial enclaves in the underdeveloped countries of the South. Paradoxically, the increasing dependence of the industrialised societies on natural resources, through the rapid spread of energy and resource-intensive production technologies, has been accompanied by the spread of the myth that increased dependence on modern technologies implies a decreased dependence on nature and natural resources This myth is supported by the introduction of a long and indirect chain of resource utilisation which leaves invisible the real material resource demands of the industrial processes.

Through this combination of resource intensity at the material level and resource indifference at the conceptual and political levels, conflicts over natural resources generated by the new pattern of resource utilisation are generally shrouded and overlooked. These conflicts become visible when resource and energy-intensive industrial technologies are challenged by communities whose survival depends on the conservation of resources threatened by destruction and overexploitation, or when the devastatingly destructive potential of some industrial technologies is demonstrated as in the Bhopal disaster.

For centuries, vital natural resources like land, water and forests had been controlled and used collectively by village communities thus ensuring a sustainable use of these renewable resources. The first radical change in resource control and the emergence of major conflicts over natural resources induced by non-local factors was associated with colonial domination of this part of the world. Colonial domination systematically transformed the common vital resources into commodities for generating profits and growth of revenues. The first industrial revolution was to a large extent supported by this transformation of commons into commodities which permitted European industries access to the resources of South Asia.

With the collapse of the international colonial structure and the establishment of sovereign countries in the region, this international conflict over natural resources was expected to be reduced and replaced by resource policies guided by comprehensive national interests. However, resource use policies continued along the colonial pattern and, in the recent past, a second drastic change in resource use has been initiated to meet the international requirements and the demands of the elites in the Third World, leading to yet another acute conflict among the diverse interests. The most seriously threatened interest, in this conflict, appears to be that of the politically weak and socially disorganised group whose resource requirements are minimal and whose survival is primarily dependent directly on the products of nature outside the market system. Recent changes in resource utilisation have almost wholly by-passed the survival needs of these groups. These changes are primarily guided by the requirements of the countries of the North and of the elites of the South. This book analyses environmental conflicts in contemporary human society. In general it relates to societies all over the world, but in particular it addresses the most intense and emerging social contradictions in India related to conflicts over natural resources. Science and technology are central to these conflicts because while scientific knowledge has been used by contemporary societies to considerably enlarge man's access to natural resources, it has also allowed the utilisation natural resources at extremely high rates.

Emergence of Ecology Movements

The contemporary period is characterised by the emergence of ecology movements in all parts of the world which are attempting to redesign the pattern and extent of natural resource utilisation to ensure social equality and ecological sustainability. Ecology movements emerging from conflicts over natural resources and the people's right to survival are spreading in regions like the Indian subcontinent where most natural resources are already being utilised to fulfil the basic survival needs of a large majority of people.

The introduction of resource and energy-intensive production technologies under such conditions leads to economic growth for a small minority while, at the same time, undermines the material basis for the survival of the large majority.

In this way, ecology movements have questioned the validity of the dominant concepts and indicators of economic development. The ideology of conomic development, which remained almost monolithic in the post World Var II period, is thus faced with a major foundational challenge.

In this chapter an attempt has been made to provide a systematic onceptual framework for analysing the processes and structures of modern conomic development from an ecological -perspective. It attempts to analyse he relationship between economic development and conflicts over natural esources to trace the roots of ecological movements.

Further, in the light of the ecological perspective, it examines the undamental assumptions and categories of modern development economics hat are used to determine the objectives of economic development as well as he criteria for the choice of technologies that are used to achieve these bjectives.

Ecology Movements and Survival

The intensity and range of ecology movements in independent India have continuously widened as predatory exploitation of natural resources to feed the process of development has increased in extent and intensity. This process has been characterised by the massive expansion of energy and resource-intensive industrial activity and major development projects like large dams, forest exploitation, mining and energy-intensive agriculture. The resource demand of development has led to the narrowing of the natural resource base for the survival of the economically poor and powerless, either by direct transfer of resources away from basic needs or by destruction of the essential ecological process that ensure renewability of the life-supporting natural resources.

In the light of this background, ecology movements emerged as the people's response to this new threat to their survival and as a demand for the ecological conservation of vital life-support systems. The most significant life-support systems in addition to clean air are the common property resources of water, forests and land on which the majority of the poor people of India depend for survival. It is the threat to these resources that has been the focus of ecology movements in the last few decades. Among the various ecology movements in India, the Chipko movement (embrace the trees to oppose fellings) is the most well known.

It began as a movement of the hill people in the state of Uttar Pradesh to save the forest resources from exploitation by contractors from outside.' It later evolved into an ecological movement that was aimed at the maintenance of the ecological stability of the major upland watersheds in India. Spontaneous people's response to save vital forest resources was seen in Jharkhand area in Bihar-Orissa border region as well as in Bastar area of Madbya Pradesh where there were attempts to convert the mixed natural forests into plantations of commercial tree species, to the complete detriment of the tribal people. In the southern part of India the Appiko movement, which was inspired by the success of the Chipko movement in the Himalayas, is actively involved in stopping illegal over-felling of forests and in replanting forest lands with multipurpose broad leaved tree species. In Himachal Pradesh the Chipko activists have intensified their opposition to the expansion of monoculture plantation of the commercial Chir Pine (Pinus roxburghii). In the Aravalli Hills of Rajasthan there has been a massive programme of tree planting to give employment to those hands which were hitherto engaged in felling of trees.

The exploitation of mineral resources, in particular the opencast mining in the sensitive watersheds of the Himalayas, the Western Ghats and Central India have also resulted in a great deal of environmental damage. As a consequence, environmental movements have come up in these regions to oppose the reckless mining operations. Most successful among them is the movement against limestone quarrying in the Doon Valley. Here, volunteers of the Chipko movement have led thousands of villagers, in peaceful resistance, to oppose the reckless functioning of limestone quarries that is seen by the people as a direct threat to their economic and physical survival.'

While the Doon Valley instance has a long history of popular opposition to the quarrying of limestone and a Supreme Court order has restricted the area of quarrying to a minimum, examples of such success' of ecology movements are rare People's ecology movements against mineral exploitation in the neighbouring areas of Almora and Pithoragarh still seem to be ignored, probably due to the relative isolation of these interior areas. Beyond the Himalayas, the ecology movement in the Gandhamardan Hills in Orissa against the ecological havoc of bauxite mining has gained momentum and it draws inspiration from the Chipko movement.

The mining project of the Bharat Aluminium Company (BALCO) in the Gandhamardan Hills is being opposed by local youth organisations and tribal people whose survival is directly under threat. The peaceful demonstrators have claimed that the project could be only continued 'over our dead bodies. The situation is more or less the same in large parts of Orissa-Madhya Pradesh region where rich mineral and coal deposits are being opened up for exploitation and thousands of people in these interior areas are being pushed to deprivation and destitution. This is also true of the coal mining areas around the energy capital of the country in Singrauli. In these interior areas of Central India, movements against both mining and forestry are becoming increasingly volatile and people's resistance is growing.

Large river valley projects, which are coming up in India at a very rapid pace, is another group of development projects against which people have organised ecology movements. The large-scale submersion of forest and agricultural lands, a prerequisite for the large river valley projects, always takes a heavy toll of dense forests and the best food growing lands. These have usually been the material basis for the survival of a large number of people in India, specially tribal people.

The Silent Valley project in Kerala was opposed by the ecology movement on the ground of its being a threat, not to the survival of the people directly, but to the gene pool of the Tropical Rainforests threatened by submersion. The ecological movement against the Tehri high dam in the UP Himalaya exposes the possible threat to people living both above and below the dam site through large-scale destabilization of land by seepage and strong seismic movements that could be induced by impoundment. The Tehri Dam Opposition Committee has appealed to the Supreme Court against the proposed dam by identifying it as a threat to the survival of all people living near the river Ganga up to West Bengal.

Most notable among the people's movements against dams on the issue of direct threat to survival from submersion are Bedthi lcchampalli, Bhopalpatnam, Narmada Sagar, Koel-Karo, Bodhghat, etc. In the context of the already overutilised land resources, the proper rehabilitation on a land-to-land basis of millions of people displaced through the construction of dams seems impossible. The cash compensation given instead is inadequate in all respects for providing an alternate livelihood for the majority of the displaced. Destitution is thus the first and foremost precondition for initiating large dam projects

Process of Construction of Dams

While the process of construction of dams itself invites opposition from ecology movements, the functioning of water projects dependent on the constructed dams results in further ecological disasters and movements. People's movements against widespread water-logging, salinisation and the resulting desertification in the command areas of many dams have been registered. Among them are instances of protests against the Tawa, Kosi, Gandak, Tungabhadra, Malaprabha, Ghatprabha projects and the canal irrigated areas of Punjab and Haryana. While excess water led to ecological destruction in these cases, improper and unsustainable use of water in the arid and semi-arid regions generated ecology movements in a different way. The anti-drought and desertification movement is gaining momentum in the dry areas of Maharashtra, Karnataka, Rajasthan, Orissa, etc. Ecological water use for survival is being advocated by water based movements like Pani Chetana, Pani Panchayat, and Mukti Sangharsh. Another major movement originating from the ecological destruction of resources by growth based development is spreading all along the 7,000 km long coastline of India. It is the movement of the small fishing communities against the ecological destruction caused by mechanised fishing whose instant profit motive is destroying the coastal ecology and its long-term biological productivity in a big way. No amount of threat to survival in India from environmental hazards can be complete without a reference to the Bhopal tragedy on 2 December 1984, in which several thousand people died and several lakhs faced serious health hazards following the leakage of poisonous Methyl Iso Cyanate from a pesticide plant of Union Carbide (India) Limited. People's movements for clean air and water are growing in ail parts of the country just as ecologically irresponsible industrialization is moving deeper into the hinterland in search of new resources.

THE SURVIVAL ECONOMY

Modern economics and the concept of development cover a miniscule portion in the history of economic production by human beings. The survival economy has given human societies the material basis of survival by deriving livelihoods directly from nature through self-provisioning mechanisms. In most Third World countries large numbers of people are deriving their sustenance in the survival economy in ways that remain invisible to market oriented development. Within the context of a limited resource base the destruction of the survival economy takes place through the diversion of natural resources from directly sustaining human existence to generating growth in the market economy. Sustenance and basic needs satisfaction is the organising principle for natural resource use in the survival economy whereas profits and capital accumulation are the organising principles for the exploitation of resources for the market. Human survival in India even today is largely dependent on the direct utilisation of common natural resources."

Ecology movements are voicing their opposition to the destruction of these vital commons so essential for human survival. Without clean water, fertile soils, and crop and plant genetic diversity economic development will become impossible. Sometimes by omission and sometimes by commission formal economic development activities have impaired the productivity of common natural resources which has enhanced the contradiction between the economy of natural processes and the survival economy.

The organising principles of economic development based on economic growth renders valueless all resources and resource processes that are not priced in the market and are not inputs to commodity production. This premise very often generates economic development programmes that divert or destroy the resource base for survival. While the diversion of resources, like diversion of land from multipurpose community forests to monoculture plantations of industrial tree species, or the destruction of common resources, or the diversion of water from staple food crops and drinking water needs to cash crops are frequently proposed as programmes for economic development in the context of the market economy, they create economic underdevelopment in the economies of nature and survival. Ecology movements are aimed at opposing these threats to survival from market based economic development. Thus in the Third World, ecology movements are not the luxury of the rich; they are a survival imperative for the majority of people whose survival is not taken care of by the market economy but is threatened by its expansion.

The political foundation of ecology movements lies in their capacity to enlarge the spatial, temporal and social bases for the evaluation of economic development projects-in their capacity to bring into the picture all the three economies described earlier. A new economics of development will emerge only when these three economies can be conceptualized within a single framework.

ECOLOGICAL CRITERIA AND ECONOMIC DEVELOPMENT PROGRAMMES

When economic development programmes are viewed from the perspective of all the three economies, a clearer view of the political economy of conflicts over natural resources is expected to emerge. In the dominant mode of economic development, perceived within the framework of the market economy, mediation of technology is assumed to lead to the control of larger and larger quantities of natural resources, thus turning scarcity into abundance and poverty into affluence: Technology, accordingly is viewed as the motive force for development and the vital instrument that guarantees freedom from dependence on nature ' The affluence of the industrialized west is assumed to be associated exclusively with this capacity of modern technology to generate wealth.

The concept of technology per se as a source of abundance and freedom from nature's ecological limits are based in part on the limitations of the market economy in understanding in a holistic manner, the same resources which it exploits. Only when development processes are viewed in the holistic perspective of all the three economies can the scarcities and underdevelopment associated with abundance and development be clearly seen. Most resource-intensive technologies operate in the enclaves with enormous amounts of various resources coming from diverse ecosystems which are normally far away. This long, indirect and spatially distributed process of resource transfer made possible by energy-intensive long distance transportation, leaves invisible the real material demands of the technological processes of development.

The spatial separation of resource exhaustion and the creation of products have also considerably shielded the inequality creating tendencies of modern technologies. Further, it is simply assumed that the benefits of economic development based on these modern technologies will automatically percolate to the poor and the needy and growth will ultimately take care of the problems of distributive justice. This would, of course, be the case, if growth and surplus were in a sense absolute and purchasing power existed in all socio-economic groups. None, however, is correct. Surplus is often generated at the cost of the ecological productivity of natural resources or at the cost of exhausting the capital of non-renewable resources. For the poor, the only impact of such economic activity often is the loss of their resource base for survival.

It is thus no accident that modern, efficient and 'productive' technologies 'creased within the context of growth in market economic terms are associated with heavy social and ecological costs. The resource and energy intensity of the production processes they give rise to demands ever increasing resource withdrawals from the natural ecosystems. These excessive withdrawals in the course of time disrupt essential ecological processes and result in the conversion of renewable resources into non-renewable ones. Over time, a forest provides inexhaustible supplies of water and biomass including wood, if its capital stock, diversity and hydrological stability are maintained and it is harvested on a sustained yield basis.

The heavy and uncontrolled market demand for industrial and commercial wood, however, requires continuous over-felling of trees which destroys the regenerative capacity of the forest ecosystems and over time converts these forests into non-renewable resources. Sometimes the damage to nature's intrinsic regenerative capacity is impaired not directly by over-exploitation of a particular resource but indirectly by damage caused to other natural resources related through ecological processes.

Thus under tropical monsoon conditions, over-felling of trees in catchment areas of streams and rivers not only destroys forest resources, but also stable, renewable sources of water. Resource-intensive industries do not merely disrupt essential ecological processes by their excessive demands for raw materials; they also destroy and disrupt vital ecological processes by polluting essential resources like air and water.

In the words of Rothman: 'the private economic rationality of the profit seeking business enterprise is a murderous providence because it cannot guarantee the optimum use of resources for society as a whole. It cannot avoid continually creating situations which cause the pollution of an environment.

In the context of resource scarcity where most resources are already being utilised for the satisfaction of survival needs, further diversion of resources to new uses is likely to threaten survival and generate conflicts between the demands of economic growth and the requirements of survival. It, therefore, becomes essential to evaluate the role of new technologies in economic development on the basis of their resource demands and conflict with the demands of survival. The productivity of 8 technology in the perspective of human survival must distinguish outputs in terms of their potential for satisfaction of vital or non-vital needs, because on the continued satisfaction of vital needs depends human survival. As Georgescu-Roegen points out.

There can be no doubt about it. Any use of the natural resources for the satisfaction of non-vital needs means a smaller quantity of life in the future. If we understand well the problem, the best use of our iron resources is to produce plows or harrows as they are needed, not Rolls Royces, not even agricultural tractors.

In the context of the market economy, the indicators of technological efficiency and productivity are totally independent of the difference between the satisfaction of basic needs and luxury requirements between resources extracted by ecologically sensitive or insensitive technologies or of the nature of the contribution of economic growth to diverse socio-economic categories. In the context of a highly non-uniform distribution of purchasing power and scanty knowledge of or respect for ecological processes, economic growth depends on production and consumption of nonvital products.

The expansion of the formal sector of the economy for the production of non-vital goods often leads to further diversion of vital natural resources.

For example, water-intensive production of flowers or fruits for the lucrative export market often results in water scarcity in low rainfall areas. In a world with a limited and shrinking resource base, and in the economic framework of a market economy, non-vital luxury needs are fulfilled at the cost of vital survival needs. The high powered pull of the purchasing capacity of the rich of the world can draw out necessary resources in spite of resource scarcity and resulting conflicts.

This complete lack of recognition of the resource needs of the survival economy nature's economy in the current paradigm of development economics shrouds the political issues arising from resource transfer and ecological destruction. For the economic sector based on 'efficient modern technologies', this provides an ideological weapon for increased control of the sponsors of economic development over the total natural resource endowments of the countries concerned.

CONCEPT OF 'PRODUCTIVITY'

The ideological and limited concept of 'productivity' of technologies has been universalised with the consequence that all other costs of the economic process become invisible. The invisible forces which contribute to the increased 'productivity' of a modern farmer or factory worker emanate from the increased consumption of non-renewable natural resources. Lovins has described this as the amount of 'slave' labour at present at work in the world. According to him, each person on earth, on an average, possesses the equivalent of about fifty slaves, each working forty hours a week. Man's annual global energy conversion from all sources (wood, fossil fuel, hydroelectric power, nuclear) at present is approximately 8 x 10 (12) watts. This is more than twenty times the energy content of the food necessary to feed the present world population at the FAO standard per capita requirement of 3,600 cals per day.

In terms of workforce, therefore, the population of the earth is not 4 billion but about 200 billion, the important point being that about 98 per cent of them do not eat conventional food. The inequalities in the distribution of this 'slave' labour between different countries is enormous, the average inhabitant of the USA, for example, having 250 times as many 'slaves' as the 'average Nigerian'. And this, substantially is the reason for the difference in efficiency between the American and Nigerian economies: it is not due to the differences in the average 'efficiency' of the people themselves. There seems no way of discovering the relative efficiencies of Americans and Nigerians: If Americans were short of 249 of every 250'slaves' they possess, who can say how 'efficient' they would prove themselves to be.

The increase in the levels of resource consumption is taken universally as an indicator of economic development. If the present level of resource consumption in the USA is accepted as the development objectives of India, the total resource demands of 'developed' India can be calculated by multiplying the current resource consumption by a factor of 250. Neither our forests nor our fields or rivers can sustain such a 'development'. When per capita resource consumption is considered, the Malthusian argument relating population with resource scarcity does not hold good. More significant than the population factor is the total resource factor. Thus, although many countries of the South have a much larger population than those of the North, the industrialized of the world consumes more grain than all the other three-quarters put together. This high consumption is due to the fact that intensive livestock production in industrialized countries accounts for 67 per cent of their total grain consumption. This Efficient' process of livestock management for the production of meat, as reported by Odium requires 10 calories of energy input to produce one calorie of food energy.

The energy subsidy provided by the capital stock of the earth's non-renewable resources makes a resource inefficient process appear as efficient in the market economy. It is interesting to note that even in the West, nearly a century ago one calorie of food was produced by using a fraction of a calorie of energy input. The same is true in the economics of water resources use in modern agriculture. When the production of high yielding varieties of seeds is evaluated, not on productivity per unit land (tons/ha) but per unit volume of water input (tons/le litre), these miracle seeds of the Green Revolution are seen as two to three times less efficient in food production than, say, the millets. The results of evaluation of the technological efficiency of processes associated with economic development, when reexamined on a holistic basis and optimised against all resource inputs, would generally lead to the conclusion that: 'the much talked of efficiency of widely practiced high technology is not intrinsically true. They are, in fact, highly wasteful of materials and pollutive (that is, destructive to the productive potential of the environment)'.

New technologies in the market economy are innovated for profit maximization and not to encourage resource prudence per se. The extent of inefficiency in the utilisation of natural resources with production processes based on resource-intensive technologies, can be illustrated with the production of soda ash, an important industrial material. In the Solvay process for the production of soda ash the two materials used are sodium chloride and limestone.

The entire limestone used in the process ends up as waste material, 25 per cent of the sodium chloride is lost as unreacted salt. From the balance 75-80 per cent, the acidic half is lost and only the basic half goes into the final product. Therefore only 40 per cent of the raw materials consumed are actually utilised. The waste products pollute land and water resources systems. The economy of the process is artificially made good by concessions in procuring limestone, salt and fuel and further concessions in respect of land, transport, etc. It is these subsidies for natural resources which make the counter-productive processes appear efficient.

Referring to the technology of production of frozen orange juice Schnaiberg made the following remarks:

What is true of the unobtrusive shift from fresh oranges to frozen orange juice is typical of most transitions from traditional to late industrial technologies. The majority of these become more energy intensive: the energy content of all the necessary production processes increases per unit produced.... The hall mark of modern technology is its typical labour saving quality-not its energy saving aspect."

Guided by a narrow and distorted concept of efficiency and supported by all types of subsidies, technological change in market economy-oriented development continues in the direction of resource intensity, labour displacement and ecological destruction. The long-term continuation of such processes will lead to the destruction of the resource base of the survival economy and to human labour being rendered dispensable in the production processes of the market economy. The partisan assumptions of modern economic development which cannot internalise the economy of natural processes and the survival economy are thus being raised to the level of universality. As a result, with the expansion of economic development in Third World countries, the resource-intensive and socially partial development is leading to social instability and conflicts. While ecology movements in the industrially advanced countries are directed against more recent threats to survival like pollution, ecology movements in Third World countries have a much longer history related to resource exhaustion and ecological degradation of natural ecosystems. It is in these countries that the holistic ecological criteria for technology choice is needed most urgently.

The process of transformation and utilisation of natural resources for the satisfaction of societal needs determines the economic organization of human societies. At various stages of development, the dominant patterns of utilisation of natural resources have been guided by the dominant pattern of scientific knowledge, and through the generation and use of technologies that actually bridge the gap between natural resources and human needs and requirements.

A special characteristic of human societies is that they can make deliberate choices between different ways of using resources and satisfying needs. The existence of plurality of alternatives in resource use for economic development creates the need for a selection criteria to make rational decisions about the use of natural resources and technological change. A dialectical relationship exists between the criteria of technology choice and the nature of science and technology developed in response to the criteria. Traditional societies as well as modern scientific-industrial societies have adopted different systems of science and technology which differ primarily in the criteria of choice or rationality that guides resource use patterns for human needs satisfaction. The characterization of certain societies as primitive and unscientific is, thus, sociologically and epistemologically unfounded. The fact that values and rationality criteria of one form of social organization generate a particular type of science and technology matched to a particular criteria of scientificity does not imply that other social organisations lack a scientific basis for their economic activities.

If sustainable utilisation is the objective that guides the criteria of choice for a development strategy, a resource prudent technological path (T.) is rationally chosen. If maximization of the growth of man-made processes and increasing the productivity of labour is the objective, then a more resource-intensive path (T2) which is the integration of a large number of smaller technologies (t1) and in which increased resource and energy inputs allows the increase in labour productivity, is rationally chosen. In this process a large amount of secondary resources (R2-R6) are additionally required.

Traditional societies in all their diversity have, in general, shared a common set of characteristics. They have used natural resources prudently to satisfy minimum needs sustainably over centuries. Such resource use was based on.

Resource flow in resource prudent t1 and resource.-intensive t2 technology chains

1. A knowledge system with an ecological understanding of nature.
2. A technological system for processing resources to satisfy human needs with minimum resource waste.
3. Rationality criteria for demarcating vital and non-vital needs and between resource destructive and resource enhancing technologies.

Traditional world views and practices deterred over-exploitation of natural resources at all levels. As they were based on ecological perceptions of nature and guided by restraints in resource use, they used technologies which prevented ecological disruption.

Modernisation of traditional societies in its present form has, by and large, been taken as synonymous with the substitution of indigenous science and technology systems by the modern western system. In this manner the resource-intensive western pattern of resource use is thrust on non-western societies through modernisation.

Modern western scientific knowledge, however, differs from indigenous knowledge systems in three important ways:

1. Modern western scientific knowledge is reductionist and fragmented.
2. Modern western technological systems are based on reductionist science and are generally more resource-intensive.
3. There are no criteria of rationality or technology choices to evaluate modern science and technology on the basis of resource use efficiency or need satisfaction capability.

These characteristics of modern western science and technology systems breaks the chain, beginning with natural resources and ending in the satisfaction of human needs and demands, into small fragments of individually identifiable economic activities. This provides justification for the resource intensity of the dominant paradigm of economic development and technological change, and thus leads to ecological instabilities.

Ecological crises are thus inevitable products of economic activities which are propelled towards longer and more complex and resource-intensive technological chains (T2) for the satisfaction of older needs (N.). Only individual segments(l) of the whole technological chain are examined from the narrow criteria of labour productivity. The situation is best exemplified in the case of food production. While indigenous and traditional food production practices used about half a calorie of energy to produce 1 calorie of food, the present mechanised and chemical farming techniques use 10 calories of energy to produce 1 calorie of food. These characteristics of contemporary scientific industrial development are the primary causes for the contemporary ecological crises. The combination of eco!ogically disruptive scientific and technological modes, and the absence of rationality criteria for evaluating scientific and technological systems in terms of resource use efficiency, has created conditions where society is increasingly propelled towards ecological instability and has no rational and organised response to arrest and curtail these destructive tendencies.

EARTH'S ENERGY BUDGET

The Earth can be considered as a physical system with an energy budget that includes all gains of incoming energy and all losses of outgoing energy. The planet is approximately in equilibrium, so the sum of the gains should be approximately equal to the sum of the losses.

Note: although the term "energy budget" is widely used, the flow of energy in and out of the Earth is actually measured in units of power (watts), not units of energy (joules). Therefore, "power budget" would be a more accurate term.

The Power Budget

Incoming Power

The total flux of power entering the Earth's atmosphere is estimated at 174 peta watts. This consists of:

- solar radiation (99.978%, or nearly 174 petawatts; or about 340 W m-2)
  - This is equal to the product of the solar constant, about 1366 watts per square metre, and the area of the Earth's disc as seen from the Sun, about 1.28 × 1014 square metres, averaged over the Earth's surface, which is four times larger. The solar flux averaged over just the sunlit half of the earth's surface is about 680 W m-2

- geothermal energy (0.013%, or about 23 terawatts; or about 0.045 W m-2)
  - This is produced by stored heat and heat produced by radioactive decay leaking out of the Earth's interior.
- tidal energy (0.002%, or about 3 terawatts; or about 0.0059 W m-2)
  - This is produced by the interaction of the Earth's mass with the gravitational fields of other bodies such as the Moon and Sun.
- waste heat from fossil fuel consumption (about 0.007%, or about 13 terawatts; or about 0.025 W m-2).

Note that the solar constant varies (by approximately 0.1% over a solar cycle); and is not known absolutely to within better than about one watt per square metre. Hence the geothermal and tidal contributions are less than the uncertainty in the solar power.

Outgoing Power

The average albedo (reflectivity) of the Earth is about 0.3, which means that 30% of the incident solar energy is reflected back into space, while 70% is absorbed by the Earth and reradiated as infrared.

The planet's albedo varies from month to month, but 0.3 is the average figure. It also varies very strongly spatially: ice sheets have a high albedo, oceans low.

The contributions from geothermal and tidal power sources are so small that they are omitted from the following calculations.

The 30% reflected energy consists of:

- 6% reflected from the atmosphere
- 20% reflected from clouds
- 4% reflected from the ground (including land, water and ice)

All of the 70% absorbed energy is eventually reradiated:

- 64% by the clouds and atmosphere
- 6% by the ground

The same 70% of absorbed energy can be split this way:

- 51% absorbed by land and water, then emerging in the following ways:
  - 23% transferred back into the atmosphere as latent heat by the evaporation of water
  - 7% transferred back into the atmosphere by heated rising air
  - 6% radiated directly into space
  - 15% transferred into the atmosphere by radiation, then reradiated into space and
- 19% absorbed by the atmosphere, including:

- 16% reradiated back into space
- 3% transferred to clouds, from where it is radiated back into space

Anthropogenic Modification

Emission of greenhouse gases, and other factors (such as land-use changes), modify the energy budget slightly but significantly. The IPCC provides an estimate of these forcing, insofar as they are known.

The largest and best known are from the well-mixed greenhouse gases (CO_2, CH_4, halocarbons, etc.), totalling an increase in forcing of 2.4 W m-2 relative to 1750.

These are less than 1% of the solar input, but contributes to the observed increase in atmospheric and oceanic temperature.

ELECTRICITY GENERATION

Electricity generation is the first process in the delivery of electricity to consumers. The other processes are electric power transmission and electricity distribution.

The importance of dependable electricity generation, transmission and distribution was revealed when it became apparent that electricity was useful for providing heat, light and power for human needs. Centralized power generation became possible when it was recognized that alternating current electric power lines can transport electricity at low costs across great distances by taking advantage of the ability to transform the voltage using power transformers.

Electricity has been generated for the purpose of powering human technologies for at least 120 years from various sources of energy. The first power plants were run on wood, while today we rely mainly on petroleum, natural gas, coal, hydroelectric and nuclear power and a small amount from hydrogen, solar energy, tidal harnesses, wind generators, and geothermal sources.

Electricity Demand

The demand for electricity can be met in two different ways. The primary method thus far has been for public or private utilities to construct large scale centralized projects to generate and transmit the electricity required to fuel economies. Many of these projects have caused unpleasant environmental effects such as air or radiation pollution and the flooding of large areas of land. Distributed generation creates power on a smaller scale at locations throughout the electricity network. Often these sites generate electricity as a byproduct of other industrial processes such as using gas from landfills to drive turbines.

Methods of Generating Electricity

Turbines: Rotating turbines attached to electrical generators produce most commercially available electricity. Turbines are driven by a fluid which acts as an intermediate energy carrier. The fluids typically used are:

- *steam*-Water is boiled by nuclear fission or the burning of fossil fuels (coal, natural gas, or petroleum). Some newer plants use the sun as the heat source: solar parabolic troughs and solar power towers concentrate sunlight to heat a heat transfer fluid, which is then used to produce steam.
- *water*-Turbine blades are acted upon by flowing water, produced by hydroelectric dams or tidal forces,
- *wind*-Most wind turbines generate electricity from naturally occurring wind. Solar updraft towers use wind that is artificially produced inside the chimney by heating it with sunlight.
- *hot gases*-Turbines are driven directly by gases produced by the combustion of natural gas or oil.

Combined cycle gas turbine plants are driven by both steam and gas. They generate power by burning natural gas in a gas turbine and use residual heat to generate additional electricity from steam. These plants offer efficiencies of up to 60%.

Reciprocating Engines

Small electricity generators are often powered by reciprocating engines burning diesel, biogas or natural gas. Diesel engines are often used for back up generation, usually at low voltages. Biogas is often combusted where it is produced, such as a landfill or wastewater treatment plant, with a reciprocating engine or a microturbine, which is a small gas turbine.

Photovoltaic Panels

Unlike the solar heat concentrators mentioned above, photovoltaic panels convert sunlight directly to electricity. Although sunlight is free and abundant, solar panels are expensive to produce and have only a 10-20% conversion efficiency. Until recently, photovoltaics were most commonly used in remote sites where there is no access to a commercial power grid, or as a supplemental electricity source for individual homes and businesses. Recent advances in manufacturing efficiency and photovoltaic technology, combined with subsidies driven by environmental concerns, have dramatically accelerated the deployment of solar panels. Installed solar capacity is growing by 30% per year in several regions including Germany, Japan, California and New Jersey.

WORLD ENERGY CONSUMPTION

The world has a number of energy resources which provide us with the ability to perform work (a narrowly-defined measure of the ability to effect physical change).

Energy resources range from fossil fuels and nuclear fuels to renewable energy such as wind, hydro-and solar energy.

When this energy is drawn upon to do work, and is converted to a less useful form, energy consumption is said to have occured.

Total Consumption

In 2004 the worldwide energy consumption of the human race was estimated as 15 TW (TW=1012 Watts) by the United States Energy Information Administration. This is equivalent to 0.5 ZJ (ZJ =1021J) per year. There is at least 10% uncertainty in this number because the underlying data is continuously changing and not all of the world's economies track their energy consumption with the same rigor.

Consumption by Fuel Type

Eighty-seven percent of the world's energy is supplied by fossil fuel. The 15 TW total energy consumption of 2004 was divided as follows:

	TW=1012 Watt	ZJ =1021J per year
Oil	5.6	0.18
Gas	3.5	0.11
Coal	3.8	0.12
Hydroelectric	0.9	0.03
Nuclear	0.9	0.03
Geothermal, Wind, Solar, Wood	0.2	0.006

Ever since the advent of the industrial revolution, the worldwide energy consumption has been growing steadily. In 1890 the consumption of fossil fuels roughly equaled the amount of bio mass fuel burned by households and industry. In 1900 global energy consumption equaled 0.7 TW. The twentieth century saw a rapid twenty fold increase in fossil fuels. Between 1980 and 2004 the worldwide annual growth rate has been 2%.

The most significant growth of energy consumption is currently taking place in China which has been growing at 5.5% over the last 25 years. Its 1.3 billion people are currently consuming energy at a rate of 2 KW per person. Over the last twenty years Fossil fuels have continued to grow and increased their share of the energy supply, coal has been growing the fastest. Nuclear and hydro electric energy have stagnated due to environmental concerns. With the exception of France no major western country has ordered a new nuclear power plant since 1980. In 2004 renewable energy supplied around 7% of the world's energy consumption. Most (87%) of the renewable energy was hydroelectric power. Solar and wind power provided 4 and 65 GW (GW=109 Watt) respectively.

Consumption by Country

Energy consumption broadly tracks with gross national product, although there is a significant difference between the consumption levels of the US with 11.4 KW per person, and Japan and Germany's with 6 KW per person. In developing countries such as India the per person energy use is closer to 0.5 KW. The per capita energy consumption against the GDP per person for all countries with more than 20 million inhabitants, this covers 95% of the world population. Canada has the highest energy consumption per capita whereas the lowest energy consumption takes place in the third world.

Energy Consumption by Sector

Transportation, industry, residential and commercial (offices & shops) each consume approximately equal shares of the total 15 TW.

A third of the world's energy is used to produce electricity. In 2005, global electricity consumption equaled 2 TW. The energy used to generate 2 TW of electricity is approximately 5 TW, as the efficiency of a typical existing power plant is around 38%. The new generation of gas fired plants reaches a substantially higher efficiency of 55%. Coal is the most popular fuel for the world's electricity plants.

WORLD ENERGY RESOURCES

Renewable Energy Resources

Political and/or environmental considerations might move the world's energy consumption away from fossil fuels. For example, the European Commission has proposed in its Renewable Energy Roadmap21, a binding target of increasing the level of renewable energy in the EU's overall mix from less than 7% today to 20% by 2020. In 2005 the Swedish government announced their intention to become the first country to break their dependence on oil and other 'fossil raw materials' by 2020.

Renewable energy sources are even larger than the traditional fossil fuels and in theory can easily supply our energy needs. Figure 6 shows how the the sunlight interacts with the planet. There is 89,000 TW of solar energy that falls on the planet's surface. In other words, we need to capture less than 0.02% of the available solar energy to meet our current energy needs.

The available wind energy estimates range from 300 to 370 TW. Using the lower estimate, just 5% of the available wind energy would supply the current worldwide energy needs. Most of this wind energy is available over open ocean. Not only does the ocean cover 71% of the planet, wind also tends to blow stronger over open water because there are less obstructions. The above percentages for solar and wind energy ignore the formidable challenges of energy distribution and storage that need to be solved to overcome the intermittent and seasonal variations of these energy sources. The numbers do however illustrate that it is technically possible to move away from fossil fuels should society decide to make the necessary investment. The required investment to change energy sources is enormous. It is estimated that the cost to replace the global infrastructure for liquid transportation fuels alone is $3-5 trillion.

Fossil Fuel Resources

Despite several voices predicting the imminent decline of fossil fuels, there are still significant reserves of all the traditional energy components. Remaining reserves of conventional fossil fuels are estimated as:

Coal	290 ZJ
Oil	57 ZJ
Gas	30 ZJ

Again significant uncertainty exists for these numbers. This uncertainty is not surprising, since the estimation for the remaining fossil fuels on the planets depends on a detailed understanding of the earth crust. This understanding is still less than perfect. While modern drilling technology makes it possible to drill wells in up to 3000 meters of water to verify the exact composition of the geology, half of the ocean has depth exceeding 3000 meters, leaving about a third of the planet beyond the reach of detailed analysis.

The make up of the estimated 57 ZJ of oil on earth. The estimations for the world's oil reserves vary from low of 8 ZJ, consisting of currently proven and recoverable reserves, to a maximum of 110 ZJ consisting of available, but not necessary recoverable reserves, and including optimistic estimates for unconventional sources such as tar sands and oil shale. Both estimates provide oil for the foreseeable future at current oil consumption rate of 0.18 ZJ per year. There is a broad consensus among scientists that we are not close to running out of fossil fuels. Coal is especially abundant and by itself can sustain the current energy consumption of the entire planet for the next 600 years.

USE OF NUCLEAR REACTIONS TO RELEASE ENERGY

Nuclear power is the controlled use of nuclear reactions to release energy for work including propulsion, heat, and the generation of electricity. Human use of nuclear power to do significant useful work is currently limited to nuclear fission and radioactive decay. Nuclear energy is produced when a fissile material, such as uranium-235 (235U), is concentrated such that nuclear fission takes place in a controlled chain reaction and creates heat-which is used to boil water, produce steam, and drive a steam turbine.

The turbine can be used for mechanical work and also to generate electricity. Nuclear power is used to power most military submarines and aircraft carriers and provides 7% of the world's energy and 15.7% of the world's electricity. The United States produces the most nuclear energy, with nuclear power providing 20% of the electricity it consumes, while France produces the highest percentage of its electrical energy from nuclear reactors-80% as of 2006. Nuclear energy policy differs between countries.

Nuclear energy uses an abundant, widely distributed fuel, and mitigates the greenhouse effect if used to replace fossil-fuel-derived electricity. International research is ongoing into various safety improvements, the use of nuclear fusion and additional uses such as the generation of hydrogen (in support of hydrogen economy schemes), for desalinating sea water, and for use in district heating systems.

Construction of nuclear power plants in the U.S. declined following the 1979 Three Mile Island accident and the 1986 disaster at Chernobyl. Lately, there has been renewed interest in nuclear energy from national governments due to economic and environmental concerns.

Other reasons for interest include the public, some notable environmentalists due to increased oil prices, new passively safe designs of plants, and the low emission rate of greenhouse gas which some governments need to meet the standards of the Kyoto Protocol. A few reactors are under construction, and several new types of reactors are planned.

The use of nuclear power is controversial because of the problem of storing radioactive waste for indefinite periods, the potential for possibly severe radioactive contamination by accident or sabotage, and the possibility that its use in some countries could lead to the proliferation of nuclear weapons. Proponents believe that these risks are small and can be further reduced by the technology in the new reactors.

They further claim that the safety record is already good when compared to other fossil-fuel plants, that it releases much less radioactive waste than coal power, and that nuclear power is a sustainable energy source.

Critics, including most major environmental groups, believe nuclear power is an uneconomic, unsound and potentially dangerous energy source, especially compared to renewable energy, and dispute whether the costs and risks can be reduced through new technology. There is concern in some countries over North Korea and Iran operating research reactors and fuel enrichment plants, since those countries refuse adequate IAEA oversight and are believed to be trying to develop nuclear weapons. North Korea admits that it is developing nuclear weapons, while the Iranian government vehemently denies the claims against Iran.

History

The first successful experiment with nuclear fission was conducted in 1938 in Berlin by the German physicists Otto Hahn, Lise Meitner and Fritz Strassmann.

During the Second World War, a number of nations embarked on crash programs to develop nuclear energy, focusing first on the development of nuclear reactors. The first self-sustaining nuclear chain reaction was obtained at the University of Chicago by Enrico Fermi on December 2, 1942, and reactors based on his research were used to produce the plutonium necessary for the "Fat Man" weapon dropped on Nagasaki, Japan. Several nations began their own construction of nuclear reactors at this point, primarily for weapons use, though research was also being conducted into their use for civilian electricity generation.

Electricity was generated for the first time by a nuclear reactor on December 20, 1951 at the EBR-I experimental fast breeder station near Arco, Idaho, which initially produced about 100 kW.

In 1952 a report by the Paley Commission (The President's Materials Policy Commission) for President Harry Truman made a "relatively pessimistic" assessment of nuclear power, and called for "aggressive research in the whole field of solar energy". A December 1953 speech by President Dwight Eisenhower, "Atoms for Peace", set the U.S. on a course of strong government support for the international use of nuclear power.

Early Years

On June 27, 1954, the world's first nuclear power plant to generate electricity for a power grid started operations at Obninsk, USSR. The reactor was graphite moderated, water cooled and had a capacity of 5 megawatts (MW). The world's first commercial nuclear power station, Calder Hall in Sellafield, England was opened in 1956, a gas-cooled Magnox reactor with an initial capacity of 50 MW (later 200 MW). The Shippingport Reactor

(Pennsylvania, 1957), a pressurized water reactor, was the first commercial nuclear generator to become operational in the United States.

In 1954, the chairman of the United States Atomic Energy Commission (forerunner of the U.S. Nuclear Regulatory Commission) talked about electricity being "too cheap to meter" in the future, often misreported as a concrete statement about nuclear power, and foresaw 1000 nuclear plants on line in the USA by the year 2000.

In 1955 the United Nations' "First Geneva Conference", then the world's largest gathering of scientists and engineers, met to explore the technology. In 1957 EURATOM was launched alongside the European Economic Community (the latter is now the European Union). The same year also saw the launch of the International Atomic Energy Agency (IAEA).

Development

Installed nuclear capacity initially rose relatively quickly, rising from less than 1 gigawatt (GW) in 1960 to 100 GW in the late 1970s, and 300 GW in the late 1980s. Since the late 1980s capacity has risen much more slowly, reaching 366 GW in 2005, primarily due to Chinese expansion of nuclear power. Between around 1970 and 1990, more than 50 GW of capacity was under construction (peaking at over 150 GW in the late 70s and early 80s)- in 2005, around 25 GW of new capacity was planned. More than two-thirds of all nuclear plants ordered after January 1970 were eventually cancelled.

During the 1970s and 1980s rising economic costs (related to vastly extended construction times largely due to regulatory delays) and falling fossil fuel prices made nuclear power plants then under construction less attractive. In the 1980s (U.S.) and 1990s (Europe), flat load growth and electricity liberalization also made the addition of large new baseload capacity unnecessary.

A general movement against nuclear power arose during the last third of the 20th century, based on the fear of a possible nuclear accident and on fears of latent radiation, and on the opposition to nuclear waste production, transport and final storage. Perceived risks on the citizens' health and safety, the 1979 accident at Three Mile Island and the 1986 Chernobyl accident played a key part in stopping new plant construction in many countries.

Austria (1978), Sweden (1980) and Italy (1987) voted in referendums to oppose or phase out nuclear power, while opposition in Ireland prevented a nuclear programme there. However, the Brookings Institution suggests that new nuclear units have not been ordered primarily for economic reasons rather than fears of accidents.

Financing for new reactors dried up when Wall Street's enthusiasm ended. Disillusionment was complete when new research discredited the claim (previously accepted as fact even by opponents) that nuclear power was still, despite all its problems, the most cost-effective source of electricity. Industry figures had omitted the factor of downtime.

During the 1980s and early 1990s, the newest and biggest U.S. plants were actually producing only half the energy they were supposed to, due to shutdowns for refueling, routine maintenance, retrofitting, and frequent minor mishaps. Since that time, the capacity factor of existing nuclear power plants has increased dramatically, and has been near 90% in the current decade.

As of 2006, the stated desire to use nuclear power for electricity generation has been suspected of being a cover for nuclear proliferation in the countries of Iran and North Korea.

Current Technology

There are two types of nuclear power in current use: The nuclear fission reactor produces heat through a controlled nuclear chain reaction in a critical mass of fissile material. All current nuclear power plants are critical fission reactors, which are the focus of this chapter. The output of fission reactors is controllable.

There are several subtypes of critical fission reactors, which can be classified as Generation I, Generation II and Generation III. All reactors will be compared to the Pressurized Water Reactor (PWR), as that is the standard modern reactor design. The difference between fast-spectrum and thermal-spectrum reactors will be covered later. In general, fast-spectrum reactors will produce less waste, and the waste they do produce will have a vastly shorter halflife, but they are more difficult to build, and more expensive to operate. Fast reactors can also be breeders, whereas thermal reactors generally cannot.

Pressurized Water Reactors (PWR)

These are reactors cooled and moderated by high pressure liquid (even at extreme temperatures) water. They are the majority of current reactors, and are generally considered the safest and most reliable technology currently in large scale deployment, although Three Mile Island is a reactor of this type. This is a thermal neutron reactor design.

Boiling Water Reactors (BWR)

These are reactors cooled and moderated by water, under slightly lower pressure. The water is allowed to boil in the reactor. The thermal efficiency of these reactors can be higher, and they can be simpler, and even potentially more stable and safe. Unfortunately, the boiling water puts more stress on many of the components, and increases the risk that radioactive water may escape in an accident.

These reactors make up a substantial percentage of modern reactors. This is a thermal neutron reactor design.

Pressurized Heavy Water Reactor (PHWR)

A Canadian design, (known as CANDU) these reactors are heavy-water-cooled and-moderated Pressurized-Water reactors. Instead of using a single large containment vessel as in a PWR, the fuel is contained in hundreds of pressure tubes. These reactors are fuelled with natural uranium and are thermal neutron reactor designs. PHWRs can be refueled while at full power, which makes them very efficient in their use of uranium (it allows for precise flux control in the core). Most PHWRs exist within Canada, but units have been sold to Argentina, China, India (pre-NPT), Pakistan (pre-NPT), Romania, and South Korea. India also operates a number of PHWR's, often termed 'CANDU-derivatives', built after the 1974 Smiling Buddha nuclear weapon test.

Reaktor Bolshoy Moshchnosti Kanalniy (RBMK)

A Soviet Union design, built to produce plutonium as well as power, the dangerous and unstable RBMKs are water cooled with a graphite moderator. RBMKs are in some respects similar to CANDU in that they are refuelable On-Load and employ a pressure tube design instead of a PWR-style pressure vessel. However, unlike CANDU they are very unstable and too large to have containment buildings. Because of this RBMK reactors are generally considered one of the most dangerous reactor designs in use. Chernobyl was an RBMK.

Gas Cooled Reactor (GCR) and Advanced Gas Cooled Reactor (AGCR)

These are generally graphite moderated and CO2 cooled. They have a high thermal efficiency compared with PWRs and an excellent safety record. There are a number of operating reactors of this design, mostly in the United Kingdom. Older designs (i.e. Magnox stations) are either shut down or will be in the near future. However, the AGCRs have an anticipated life of a further 10 to 20 years. This is a thermal neutron reactor design.

Super Critical Water-cooled Reactor (SCWR)

This is a theoretical reactor design that is part of the Gen-IV reactor project. It combines higher efficiency than a GCR with the safety of a PWR, though it is perhaps more technically challenging than either.

The water is pressurized and heated past its critical point, until there is no difference between the liquid and gas states. An SCWR is similar to a BWR, except there is no boiling (as the water is critical), and the thermal efficiency is higher as the water behaves more like a classical gas. This is an epithermal neutron reactor design.

Liquid Metal Fast Breeder Reactor (LMFBR)

This is a reactor design that is cooled by liquid metal, totally unmoderated, and produces more fuel than it consumes. These reactors can function much like a PWR in terms of efficiency, and do not require much high pressure containment, as the liquid metal does not need to be kept at high pressure, even at very high temperatures. Superphénix in France was a reactor of this type, as was Fermi-I in the United States. The Monju reactor in Japan suffered a sodium leak in 1995 and is approved for restart in 2008. All three use/used liquid sodium. These reactors are fast neutron, not thermal neutron designs. These reactors come in two types:

Lead Cooled

Using lead as the liquid metal provides excellent radiation shielding, and allows for operation at very high temperatures. Also, lead is (mostly) transparent to neutrons, so fewer neutrons are lost in the coolant, and the coolant does not become radioactive. Unlike sodium, lead is mostly inert, so there is less risk of explosion or accident, but such large quantities of lead may be problematic from toxicology and disposal points of view. Often a reactor of this type would use a lead-bismuth eutectic mixture. In this case, the bismuth would present some minor radiation problems, as it is not quite as transparent to neutrons, and can be transmuted to a radioactive isotope more readily than lead.

Sodium Cooled

Most LMFBRs are of this type. The sodium is relatively easy to obtain and work with, and it also manages to actually remove corrosion on the various reactor parts immersed in it. However, sodium explodes violently when exposed to water, so care must be taken, but such explosions wouldn't be vastly more violent than (for example) a leak of superheated fluid from a SCWR or PWR. The radioisotope thermoelectric generator produces heat through passive radioactive decay.

Some radioisotope thermoelectric generators have been created to power space probes (for example, the Cassini probe), some lighthouses in the former Soviet Union, and some pacemakers. The heat output of these generators diminishes with time; the heat is converted to electricity utilising the thermoelectric effect.

How it Works

The key components common to most types of nuclear power plants are:

- Nuclear fuel
- Neutron moderator
- Coolant
- Control rods
- Pressure vessel
- Emergency core cooling systems
- Reactor protective system
- Steam generators (not in BWRs)
- Containment building
- Boiler feedwater pump
- Turbine
- Electrical generator
- Condenser

Conventional thermal power plants all have a heat source. Examples are gas, coal, or oil. For a nuclear power plant, this heat is provided by nuclear fission inside the nuclear reactor. When a relatively large fissile atomic nucleus (usually uranium-235 or plutonium-239) is struck by a neutron it forms two or more smaller nuclei as fission products, releasing energy and neutrons in a process called nuclear fission.

The neutrons then trigger further fission. And so on. When this nuclear chain reaction is controlled, the energy released can be used to heat water, produce steam and drive a turbine that generates electricity.

It should be noted that a nuclear explosive involves an uncontrolled chain reaction, and the rate of fission in a reactor is not capable of reaching sufficient levels to trigger a nuclear explosion because commercial reactor grade nuclear fuel is not enriched to a high enough level. (see enriched uranium) The chain reaction is controlled through the use of materials that absorb and moderate neutrons. In uranium-fueled reactors, neutrons must be moderated (slowed down) because slow neutrons are more likely to cause fission when colliding with a uranium-235 nucleus. Light water reactors use ordinary water to moderate and cool the reactors.

When at operating temperatures if the temperature of the water increases, its density drops, and fewer neutrons passing through it are slowed enough to trigger further reactions. That negative feedback stabilizes the reaction rate.

SPATIAL RELATIONSHIPS AND SUBDIVISIONS OF LAND

Ecosystems are not isolated from each other, but are interrelated. For example, water may circulate between ecosystems by the means of a river or ocean current. Water itself, as a liquid medium, even defines ecosystems. Some species, such as salmon or freshwater eels move between marine systems and fresh-water systems. These relationships between the ecosystems lead to the concept of a biome. A biome is a homogeneous ecological formation that exists over a large region as tundra or steppes. The biosphere comprises all of the Earth's biomes -- the entirety of places where life is possible -- from the highest mountains to the depths of the oceans.

Biomes correspond rather well to subdivisions distributed along the latitudes, from the equator towards the poles, with differences based on to the physical environment (for example, oceans or mountain ranges) and to the climate. Their variation is generally related to the distribution of species according to their ability to tolerate temperature and/or dryness. For example, one may find photosynthetic algae only in the photic part of the ocean (where light penetrates), while conifers are mostly found in mountains.

Though this is a simplification of more complicated scheme, latitude and altitude approximate a good representation of the distribution of biodiversity within the biosphere. Very generally, the richness of biodiversity (as well for animal than plant species) is decreasing most rapidly near the equator and less rapidly as one approaches the poles. The biosphere may also be divided into ecozones, which are very well defined today and primarily follow the continental borders. The ecozones are themselves divided into ecoregions, though there is not agreement on their limits.

Ecosystem Productivity

In an ecosystem, the connections between species are generally related to food and their role in the food chain. There are three categories of organisms:

- Producers -- usually plants which are capable of photosynthesis but could be other organisms such as bacteria around ocean vents that are capable of chemosynthesis.
- Consumers -- animals, which can be primary consumers (herbivorous), or secondary or tertiary consumers (carnivorous and omnivores).
- Decomposers -- bacteria, mushrooms which degrade organic matter of all categories, and restore minerals to the environment. And decomposers can also decompose decaying animals

These relations form sequences, in which each individual consumes the preceding one and is consumed by the one following, in what are called food chains or food network. In a food network, there will be fewer organisms at each level as one follows the links of the network up the chain.

These concepts lead to the idea of biomass (the total living matter in a given place), of primary productivity (the increase in the mass of plants during a given time) and of secondary productivity (the living matter produced by consumers and the decomposers in a given time).

These two last ideas are key, since they make it possible to evaluate the load capacity -- the number of organisms which can be supported by a given ecosystem. In any food network, the energy contained in the level of the producers is not completely transferred to the consumers. And the higher one goes up the chain, the more energy and resources is lost and consumed. Thus, from an energy-and environmental-point of view, it is more efficient for humans to be primary consumers (to subsist from vegetables, grains, legumes, fruit, etc.) than as secondary consumers (from eating herbivores, omnivores, or their products, such as milk, chickens, cattle, sheep, etc.) and still more so than as a tertiary consumer (from consuming carnivores, omnivores, or their products, such as fur, pigs, snakes, alligators, etc.). An ecosystem(s) is unstable when the load capacity is overrun and is especially unstable when a population doesn't have an ecological niche and overconsumers.

The productivity of ecosystems is sometimes estimated by comparing three types of land-based ecosystems and the total of aquatic ecosystems:

- The forests (1/3 of the Earth's land area) contain dense biomasses and are very productive. The total production of the world's forests corresponds to half of the primary production.
- Savannas, meadows, and marshes (1/3 of the Earth's land area) contain less dense biomasses, but are productive. These ecosystems represent the major part of what humans depend on for food.
- Extreme ecosystems in the areas with more extreme climates -- deserts and semi-deserts, tundra, alpine meadows, and steppes -- (1/3 of the Earth's land area) have very sparse biomasses and low productivity
- Finally, the marine and fresh water ecosystems (3/4 of Earth's surface) contain very sparse biomasses (apart from the coastal zones).

Humanity's actions over the last few centuries have seriously reduced the amount of the Earth covered by forests (deforestation), and have increased agro-ecosystems (agriculture). In recent decades, an increase in the areas occupied by extreme ecosystems has occurred (desertification).

ECOLOGICAL CRISIS AND LOSS OF ADAPTIVE CAPACITY

Generally, an ecological crisis occurs with the loss of adaptive capacity when the resilience of an environment or of a species or a population evolves in a way unfavourable to coping with perturbations that interfere with that ecosystem, landscape or species survival. It may be that the environment quality degrades compared to the species needs, after a change in an abiotic ecological factor (for example, an increase of temperature, less significant rainfalls). It may be that the environment becomes unfavourable for the survival of a species (or a population) due to an increased pressure of predation (for example overfishing). Lastly, it may be that the situation becomes unfavourable to the quality of life of the species (or the population) due to a rise in the number of individuals (overpopulation).

Ecological crises vary in length and severity, occurring within a few months or taking as long as a few million years. They can also be of natural or anthropic origin. They may relate to one unique species or to many species, as in an Extinction event. Lastly, an ecological crisis may be local (as an oil spill) or global (a rise in the sea level due to global warming).

According to its degree of endemism, a local crisis will have more or less significant consequences, from the death of many individuals to the total extinction of a species. Whatever its origin, disappearance of one or several species often will involve a rupture in the food chain, further impacting the survival of other species.

In the case of a global crisis, the consequences can be much more significant; some extinction events showed the disappearance of more than 90% of existing species at that time. However, it should be noted that the disappearance of certain species, such as the dinosaurs, by freeing an ecological niche, allowed the development and the diversification of the mammals. An ecological crisis thus paradoxically favoured biodiversity.

Sometimes, an ecological crisis can be a specific and reversible phenomenon at the ecosystem scale. But more generally, the crises impact will last. Indeed, it rather is a connected series of events, that occur till a final point. From this stage, no return to the previous stable state is possible, and a new stable state will be set up gradually. Lastly, if an ecological crisis can cause extinction, it can also more simply reduce the quality of life of the remaining individuals.

Thus, even if the diversity of the human population is sometimes considered threatened, few people envision human disappearance at short span. However, epidemic diseases, famines, impact on health of reduction of air quality, food crises, reduction of living space, accumulation of toxic or non degradable wastes, threats on keystone species (great apes, panda, whales) are also factors influencing the well-being of people.

Due to the increases in technology and a rapidly increasing population, humans have more influence on their own environment than any other ecosystem engineer.

Some common examples of ecological crises are:

- The Exxon Valdez oil spill off the coast of Alaska in 1989
- Permian-Triassic extinction event 250 million of years ago
- Cretaceous-Tertiary extinction event 65 million years ago
- Global warming related to the Greenhouse effect. Warming could involve flooding of the Asian deltas, multiplication of extreme weather phenomena and changes in the nature and quantity of the food resources.
- Ozone layer hole issue
- Deforestation and desertification, with disappearance of many species.
- Volcanic eruptions such as Mount St. Helens and the Tunguska and other impact events
- The nuclear meltdown at Chernobyl in 1986 caused the death of many people and animals from cancer, and caused mutations in a large number of animals and people. The area around the plant is now abandoned by humans because of the large amount of radiation generated by the meltdown. Twenty years after the accident, the animals have returned.

NEW TECHNOLOGY ASSESSMENT IN ECOLOGICAL CRITERIA

The resource-intensive nature of new technologies and the lack of recognition of the renewability of natural resources are, thus, at the root of the contemporary ecological crises. Ecological development as opposed to short-term economic growth, has to be based on a technological choice for the most productive means of sustainable resource utilisation This process of technological choice through the assessment of the material costs and benefits of an economic activity constitutes an ecological audit.

It differs from the conventional benefit-cost analysis in two ways. First, it evaluates benefits and costs in material terms and not in narrow financial terms based on market factors. Second, since the ecosystems perspective recognises that resources may play multifunctional roles and can have conflicting utilities. An ecological audit also takes into account which social groups and sectors will gain and which will lose materially as a result of a particular utilisation of a resource. An ecological audit also differs fundamentally from environmental impact assessments carried out in a reductionist paradigm, which does take the environment into account but merely as a bundle of fragmented and unrelated resources, as a set and not a system of resources.

Such a fragmentary approach to the environment fails to assign economic values to essential ecological processes which arise from resource linkages and which it is incapable of perceiving. This fragmentary view has led to the impression of conservation being anti development and ecology being a luxury. Piecemeal environmental solutions provided by such a fragmentary approach are incapable of offering a lasting solution to the problems of natural resource utilisation and ecological crisis. Ecological audit is, therefore, the only scientifically adequate and socially just basis for the planning and assessment of the total environmental impact of a particular economic activity.

The information for an ecological audit is provided by the ecological sciences. Contrary to the common misunderstanding that ecological concern is opposed to scientific and technological advancement, an ecological audit challenges sciences and these challenges are far greater than the ones presented by modern strategies of economic development. The economic objectives of ecological development being:

1. Satisfaction of all basic needs.
2. Economic development with sustainability.
3. Equal distribution of the costs of development.

The scientific and technological challenges posed by ecological development may be classified as follows:

1. Increased resource use efficiency for the satisfaction of basic needs of the people, e.g., development of more efficient firewood stoves, land use for high nutritional output with low resource demands.
2. Increasing GNP(2), the productivity of nature, e.g., development of more sound hydrological practices that increase infiltration and percolation to underground water resource.
3. Increasing GNP(l), the productivity of man-made processes without reducing GNP(2), the productivity of nature; e.g., improved knowledge of organic farming.

These objectives pose problems to science which are more diverse and complex in nature than those posed by sectoral growth. The methodology of scientific and technical research for ecological development accordingly, will have to be geared to this diversity and variation. At one level it will mean interdisciplinary knowledge generation without any loss in the level of sophistication and systematisation. At another level it will imply learning from the wisdom of the people who are closest to nature and who are custodians of our ecological heritage-farmers, the traditional fisherfolk, tribal people, etc.- and decanting it as public interest science, which, together with an expert knowledge of the discipline will form the knowledge base for ecological development and utilisation of natural resources.

Ecological sciences are providing a new paradigm in which the criteria of scientificity of modern science will not be strictly applicable due to its fragmented nature. Technologies will have to be evaluated in the background of not only one part of the chain of process from natural resources to the final product, but the entire technological chain. At the same time, appropriateness of technologies may not necessarily and blindly be associated with the lack of systematization that is normally associated with modern western science.

People's involvement in the evolution of ecological sciences is imperative on two counts. First, the marginalised majority have a right to determine their path of development. Second, it is the marginalised communities who retain ecological perceptions of nature at a time when the more privileged groups have lost them. Forestry science needed women of Garhwal and tribal people to remind it that catchment forests were not mines of timber but a source of water. Scientists, technologists and decision-makers need to develop a new respect for these other sciences and scientists. In the recognition of their insights, visions and day to day experiences lies the only hope for the growth of alternate ecological sciences and hence, the survival of people. The ecological perceptions of nature have been presented from outside the reductionist partisan expertise. They have emerged from the ecological perspective of the people whose survival depends on those ecological functions of natural resources which reductionist and vested interests have ignored.

The evolution of ecological knowledge in general, will depend on people's actions and movements because reductionist expertise is epistemologically and politically constrained from evolving into a non-reductionist framework. According to Feyeraband, this dynamics of the evolution of knowledge from an expert dominated to a people dominated process is the only route to a free society:

In a free society intellectuals are just one tradition. They have no special right and their views are of no special interest (except, of course, to themselves). Problems are solved not by specialists (though their advice will not be disregarded) but by the people concerned, in accordance with the ideas they value and by the procedures they regard as most appropriate... This is how the efforts of special groups combining flexibility and respect for all traditions will gradually erode the narrow and self-servicing rationalism of those who are now using tax money to destroy the traditions of the tax payers, to ruin their minds, rape their environment and quite generally turn living human beings into well trained slaves of their own barren vision of life.

The evolution of public interest-oriented ecological knowledge is, however, likely to be opposed by the reductionist partisan expertise because this 'threatens their role in society just as the enlightenment once threatened the existence of priests and theologians'.

The evolution of the ecological, sustainable and equitable utilisation of natural resources in an alternative development strategy will also, quite obviously, be opposed by the vested interests who benefit from the existing reductionist, unsustainable and inequitable utilisation pattern.

This process has already been initiated in countries like India. At one level, people's attempts at redefining development through sustainability and justice are resisted by the introduction of a false dichotomy between 'development' and 'ecology', which conceals the real dichotomy between ecological development and unsustainable economic growth. At another level, the resistance is a consequence of the rejection of peoples perception of ecological destruction as 'unscientific', 'unproved' and 'unverified'. These attempts of experts and vested interests will work against human knowledge and public interest science, and fin turn against the possibilities of human survival.

The growing conflict between the profitability imperative and the survival imperative will lead to the emergence of a politics of knowledge. It is in this sense that ecology as the foundation of an alternative public interest science and technology converges with ecology as a foundation for the politics of survival of the people.

Alternative science and technology are not utopian dreams to be kept frozen for some post-revolutionary era. As public interest science, they are emerging here and now, as an essential part of the struggle for life through the politics of ecology.

Ecology of Global Climate Change, Environment and Biodiversity

INTRODUCTION

The assumption that "culture has triumphed over nature," is mistaken, and characterizes an outdated nature-culture dualism. While in Anthropological human evolution textbooks the first part of the story is couched in evolutionary and environmental terms, the second part denies the environment a meaningful role in human history. Instead values, beliefs and issues, history, and culture constitute the key elements of the explanatory framework. This also reflected in the disciplinary separation of archeologist/ physical anthropologists versus sociocultural anthropologists: neither acknowledges their mutual reliance.

Few efforts have been made that incorporate information about how humans have altered the environment or about how environmental change revised human activity. Examples of such changes are subsistence strategies, demographical patterns, and perceptions.

To achieve this, there exists a need to develop a multidisciplinary framework. Multidisciplinarity in science is, and has been difficult to establish (Snow). Anthropology plays an important role in the development of such an framework. Its current perspective is integrative and comparative; inclusive of temporal, spatial and cultural dimensions; and dynamic. It motivates an historical focus on the dynamics of change.

GLOBAL PERMAFROST CARBON CYCLE

The Permafrost Carbon Cycle is a sub-cycle of the larger global Carbon Cycle. Permafrost is defined as subsurface material that remains below 0° C for at least two consecutive years. Because permafrost soils remain frozen for long periods of time, they store large amounts of carbon and other nutrients within their frozen framework during that time. Permafrost represents a large carbon reservoir that is seldom considered when determining global terrestrial carbon reservoirs. Recent and ongoing scientific research however, is changing this view.

The permafrost carbon cycle deals with the transfer of carbon from permafrost soils to terrestrial vegetation and microbes, to the atmosphere, back to vegetation, and finally back to permafrost soils through burial and sedimentation due to cryogenic processes.

Some of this carbon is transferred to the ocean and other portions of the globe through the global carbon cycle.

The cycle includes the exchange of carbon dioxide and methane between terrestrial components and the atmosphere, as well as the transfer of carbon between land and water as methane, dissolved organic carbon, dissolved inorganic carbon, particulate inorganic carbon and particulate organic carbon.

Soils, in general, are the largest reservoirs of carbon in terrestrial ecosystems.

This is also true for soils in the Arctic that are underlain by permafrost. Determining carbon stocks in cryosols was completed using the Northern and Mid Latitudes Soil Database.

Permafrost affected soils cover nearly 9% of the earth's land area, yet store between 25 and 50% of the soil organic carbon. These estimates show that permafrost soils are an important carbon pool. These soils not only contain large amounts of carbon, but also sequester carbon through cryoturbation and cryogenic processes.

Carbon is not produced by permafrost. Organic carbon derived from terrestrial vegetation must be incorporated into the soil column and subsequently be incorporated into permafrost to be effectively stored. Because permafrost responds to climate changes slowly, carbon storage removes carbon from the atmosphere for long periods of time. Radiocarbon dating techniques reveal that carbon within permafrost is often thousands of years old.

Carbon storage in permafrost is the result of two primary processes:

- The first process that captures carbon and stores it is syngenetic permafrost growth. This process is the result of a constant active layer thickness and energy exchange between permafrost, active layer, biosphere, and atmosphere, resulting in the vertical increase of the soil surface elevation. This aggradation of soil is the result of aeolian or fluvial sedimentation and/or peat formation. Peat accumulation rates are as high as 0.5mm/yr while sedimentation may cause a rise of 0.7mm/yr. Thick silt deposits resulting from abundant loess deposition during the last glacial maximum form thick carbon-rich soils known as yedoma. As this process occurs, the organic and mineral soil that is deposited is incorporated into the permafrost as the permafrost surface rises.

- The second process responsible for storing carbon is cryoturbation, the mixing of soil due to freeze-thaw cycles. Cryoturbation moves carbon from the surface to depths within the soil profile. Frost heaving is the most common form of cryoturbation. Eventually, carbon that originates at the surface moves deep enough into the active layer to be incorporated into permafrost. When cryoturbation and the deposition of sediments act together, carbon storage rates increase.

The amount of carbon stored in permafrost soils is poorly understood. Current research activities seek to better understand the carbon content of soils throughout the soil column. Recent studies estimate that northern circumpolar permafrost soil carbon content equals approximately 1672 Pg. This estimation of the amount of carbon stored in permafrost soils is more than double the amount currently in the atmosphere. This most recent assessment of carbon content in permafrost soils breaks the soil column into three horizons, 0–30 cm, 0–100 cm, and 1–300 cm. The uppermost horizon, 0–30 cm contains approximately 191 Pg of organic carbon. The 0–100 cm horizon contains an estimated 496 Pg of organic carbon, and the 0–300 cm horizon contains an estimated 1024 Pg of organic carbon.

These estimates more than doubled the previously known carbon pools in permafrost soils. Additional carbon stocks exist in yedoma, carbon rich loess deposits found throughout Siberia and isolated regions of North America, and deltaic deposits throughout the Arctic. These deposits are generally deeper than the 3 m investigated in traditional studies. Many concerns arise because of the large amount of carbon stored in permafrost soils. Until recently, the amount of carbon present in permafrost was not taken into account in climate models and global carbon budgets. Thawing permafrost may release great quantities of old carbon stored in permafrost to the atmosphere. Carbon stored within arctic soils and permafrost is susceptible to release due to several different mechanisms. Carbon that is stored in permafrost is released back into the atmosphere as either carbon dioxide, CO_2, or methane, CH_4. Aerobic respiration releases carbon dioxide, while anaerobic respiration releases methane.

- Microbial activity releases carbon through respiration. Increased microbial decomposition due to warming conditions is believed to be major source of carbon to the atmosphere. The rate of microbial decomposition within organic soils, including thawed permafrost, depends on environmental controls. These controls include soil temperature, moisture availability, nutrient availability, and oxygen availability.

- Methane clathrate, or hydrates, occur within and below permafrost soils. Because of the low permeability of permafrost soils, methane gas is unable to migrate vertically through the soil column. As permafrost temperature increases, permeability also increases, allowing once trapped methane gas to move vertically and escape. Dissociation of gas hydrates is common along the Arctic coastline yet, estimates for dissociation of gas hydrates from terrestrial permafrost remains unclear.
- Thermokarst/permafrost degradation as a result of climate change and increased mean annual air temperatures throughout the Arctic threatens to release large quantities of carbon back into the atmosphere. The spatial extent of permafrost decreases in warming climate, releasing large amounts of stored carbon.
- As air and permafrost temperatures change, above ground vegetation also changes. Increasing temperatures facilitate the transfer of soil carbon to growing vegetation on the surface. This transfer removes carbon from the soil and relocates it to the terrestrial carbon pool where plants process, store, and respire it, moving it to the atmosphere.
- Forest fires in the boreal forests and tundra fires alter the landscape and release large quantities of stored organic carbon into the atmosphere through combustion. As these fires burn, they remove organic matter from the surface. Removal of the protective organic mat that insulates the soil exposes the underlying soil and permafrost to increased solar radiation which in turn increases the soil temperature, active layer thickness, and changes soil moisture. Changes in the soil moisture and saturation alter the ratio of oxic to anoxic decomposition within the soil.
- Hydrologic processes remove and mobilize carbon, carrying it downstream. Mobilization occurs due to leaching, litter fall, and erosion. Mobilization is believed to be primarily due to increased primary production in the Arctic resulting in increased leaf litter entering streams and increasing the dissolved organic carbon content of the stream. Leaching of soil organic carbon from permafrost soils is also accelerated by warming climate and by erosion along river and stream banks freeing the carbon from the previously frozen soil.

Carbon is continually cycling between soils, vegetation, and the atmosphere. Currently, carbon flux from permafrost soils is minimal, however studies suggest that future warming an permafrost degradation will increase the CO_2 flux from the soils. Thaw deepens the active layer, exposing old carbon that has been in storage for decades, to centuries, to millennia. The amount of carbon that will be released from warming conditions depends on depth of thaw, carbon content within the thawed soil, and physical changes to the environment. The likelihood of the entire carbon pool mobilizing and entering the atmosphere is low despite the large volumes stored in the soil. Although temperatures are projected to rise, it does not imply complete loss of permafrost and mobilization of the entire carbon pool. Much of the ground underlain by permafrost will remain frozen even if warming temperatures increase the thaw depth or increase thermokarsting and permafrost degradation.

Warmer conditions are expected to cause spatial declines in permafrost extent and thickening of the active layer. This decline in the extent and volume of permafrost enables the mobilization of stored soil organic carbon to the biosphere and atmosphere as carbon dioxide and methane.

Additionally, these changes are believed to impact ecosystems and alter the vegetation that is present on the surface. Increased carbon uptake by plants is expected to be relatively small when compared to the amount of carbon released by permafrost degradation.

Tundra vegetation contains 0.4 kg of carbon per m^2 while a shift to boreal forests could increase the above ground carbon pool to 5 kg of carbon per m^2. Tundra soil however, contains ten times that amount.

Additionally, a sudden and steady release of carbon dioxide and methane from permafrost soils may lead to a positive feedback cycle where warming releases carbon dioxide into the atmosphere. This carbon dioxide, a greenhouse gas, causes atmospheric concentrations to increase, causing subsequent warming. This scenario is thought to be a potential runaway climate change scenario.

DEFINING AMAZONIAN DARK EARTHS

The diversity of approaches for characterizing and identifying ADE are highlighted in this volume. The authors concur that ADE should not be defined too narrowly because of the rich variation within and between ADE (Kämpf et al., 2003). The lack of a single definition or clearly defined suite of characteristics for identifying ADE is obviously frustrating for some of the participants in this volume. At this point in the study, flexible definitions that recognize the variation are healthy. How is ADE soil identified? The principle criterion is dark color. Various issues can be raised about color. How dark do soils have to be in order to be identified as ADE? Is a simple threshold on a Munsell color chart reading sufficient?

Could soils be of radically different colours but have the same composition and formation process?

What "natural" soil benchmark is used for comparison? At what point does archaeological soil (soil of archaeological settlements, monuments, middens, and earthworks) become ADE? Are ADE without the presence of archaeological artifacts anthropogenic?

The authors of this volume grapple with these questions and provide some answers. In an attempt to address the issue of continuous distribution of soil colour from brown or gray to black, scholars working in the central Amazon Region (Sombroek, 1966) introduced the term terra mulata ("brown soils") for the large transitional zone around terra preta (the classic ADE).

The diverse meanings of ADE and potential problems of communicating between scholars of different disciplines were clearly represented in the 2002 TPA Workshop and in this volume (Kämpf et al., 2003). By relying on a vague definition of ADE, we potentially open ourselves to charges of over-or underestimating the geographic extent and importance of ADE. On the other hand, a overly narrow definition might ignore the rich variation of ADE reported in this volume. Kampf et al. (2003) consider the dynamic, historical and variable nature of ADE as a subset of general anthrosols (soils produced through human activities) for their Archaeo-pedological Classification. Their new classification is an attempt to combine insights from various disciplines to address the variability and continuous variation of ADE.

Archaeologists, who identify and map ADE, tend to rely on the discipline's soil classifications and interpretations that rarely agree with those defined by modern soil science. Archaeologists are experts at recognition and interpretation of a wide range of anthropogenic soils associated with human activities on sites and landscapes. Soil scientists focus on the horizons in the profile and consider anthropogenic features as noise, perturbation and disturbance. In contrast, archaeologists define the visually and/or texturally obvious anthropogenic features in the profile and treat the horizon formation as noise, perturbation, and disturbance.

To the archaeologist, "natural" soils are only interesting in terms of defining the boundaries of the anthropogenic soils (i.e. site boundaries; sterile boundary under site, and so forth). Archaeologists are most likely to focus on the internal variation of ADE: faint patterns, changes of texture, colour, context, fill, and features to extract function and meaning. Without the internal heterogeneity within a site, we could not do archaeology. Soil scientists are less concerned with these soil nuances and attempt to characterize features representative of larger spatial areas.

The approach proposed by Kämpf et al. (2003), and to a certain degree, traditional archaeology dedicated to building chronologies, focuses on profile descriptions of small, often deep, excavations through ADE (or sites) which emphases vertical continuity and disjuncture. Since the 1970s, archaeologists have used large areal excavations often combined with sampling to recover spatial patterning of human activities and lifeways throughout the site, emphasizing a horizontal perspective. Kämpf et al. (2003) discuss the contrasting approaches used by archaeologist and soil scientists in regard to ADE and highlight the benefits of combining both approaches. ADE research can benefit from both the general chronological approach and spatial morphology of settlement and agriculture approach.

The practice of traditional archaeology is framed within the site concept (e.g. critiques by Dunnell, 1992; Fotiadis, 1992). A site is a basic discrete unit of analysis defined by concentrations of artifacts indicating settlement or other activity assumed to reflect human behaviour. I argue that adherence to the "site concept" limits our understanding of historical ecology in the Amazon Basin. The concept of landscape within Historical Ecology and archaeology of landscape are powerful alternatives to site-based approaches and are what links archaeology to historical ecology.

Rather than focus on arbitrarily defined sites, landscape approaches try to understand human activities that occur between traditional sites and across larger areas at multiple scales. In this perspective, human activity is viewed as continuous over the landscape rather than spatially contained within traditional sites. Despite new innovative methods for archaeological survey, recovery of human residues, and "non-site" landscape approaches for defining between-site human activities (e.g. Stahl, 1995), Amazonian archaeologists are drawn to "sites," usually pre-Columbian settlements, defined by conspicuous surface concentrations of pottery, lithics, and charcoal (the most commonly preserved archaeological materials).

By extension, ADE research has adopted the site concept. Are ADE discrete spatial units of analysis as presented in this volume? How can a typical black earth site be measured if it has no clearly defined boundaries or edges? Most archaeologists and historical ecologists now recognize that the earth's surface is covered with continuous distribution of artifacts and evidence of human transformation of the landscape, making it difficult, if not impossible to clearly define boundaries. For example in the agroforestry literature on the Amazon Basin, every landscape has been transformed to some degree by thousands of years of human activities (burning, selection for economic species, weeding, and artificial disturbances).

The contributors of this volume often contrast ADE with the surrounding forest soils based on the assumption that the ADE are anthropogenic and the forest is "natural". What if the benchmark against which ADE is identified and defined is also anthropogenic? If the entire Amazon Basin is to some degree anthropogenic as some historical ecologists argue, the possibility of finding a totally pristine natural soil in Amazônia after thousands of years of human disturbance for comparison as a benchmark is unlikely. The concept of domestication of landscape may provide an alternative to the site concept.

Most ADE discussed in this volume are entire sites or a subset of traditionally defined sites (most covering hectares). The boundaries of an archaeological site and its ADE do not always correspond such as in the case of the Açutuba site on the Rio Negro where surface artifact scatters much larger than the ADE (Heckenberger et al., 1999) or the Araracuara sites on the Caquetá river where ADE in the form of terra mulata extends far beyond the distribution of artifacts.

How big does a black earth footprint have to be in order to be called ADE. Many archaeological occupation sites have discrete middens that meet the content criteria of ADE; but they are of small-scale contexts within the larger archaeological site (e.g. an individual garbage pit, lens of midden on an abandoned house floor, or post holes packed with dark midden).

In addition to colour, Kern et al. (2003) stress that ADE have a greater depth of anthropogenic A horizon than typical forest soils (30-60 cm vs 10-15 cm). Although the cultural strata of most archaeological sites correspond to the modern A Horizons; there are many exceptions such as those that are deeply buried paleosols or are so thick that post-abandonment soil formation processes have not created a deep A Horizon.

ECOLOGY WITHIN BIOLOGICAL ANTHROPOLOGY

Biological anthropologists moved from 19th century and early 20th century typological approaches and race studies to the understanding of humans and their evolution via modern ideas about adaptation to the environment as a basis for understanding human variation. Ecological theory was tied also to evolutionary processes, but more in the realm of biobehavioral evolution and associated with the Darwinian concepts of "selection" and "adaptation to the environment" (Warren, 1951; Weiner, 1964). Ecological theory in biological anthropology became a fusion of evolutionary and ecological theory (Bates, 1953, 1960), along with ideas from environmental physiology (Dill et al., 1964), biogeography and human biogeography (Coon et al., 1950), demography (Spuhler, 1959), and human biology. Recently, socioecological theory has successfully been used to extend the synthesis between ecology and anthropology, focusing primarily on individuals rather than higher levels of organization. Prominent in nearly all ecological theory in anthropology has been the concept of adaptation to the environment. Ecological studies in biological anthropology were stimulated in the 1960s by the work being done in ecosystems ecology by the scientists in the International Biological Program or IBP. At this time, there were several Human Adaptability Projects associated with the IBP that were influenced by systems science and efforts to modeling complex ecological systems.

Two Early Studies

There were two original anthropological studies that were done in the 1960s that were subject to a great deal of criticism in the anthropological literature. The first is the ecological study of the Tsembaga Maring of New Guinea by the late Roy Rappaport. The work was done in the central highlands of New Guinea. The second is the energy-flow study of Andean Quechua Indians of Peru by Brooke Thomas. This work was done on the Peuvian altiplano at a base elevation of 4,000 meters above sea level. In the first case, Rappaport was a single investigator with an overwhelming task that he set out for himself. This is a long-established tradition in anthropology for one investigator to live with a people and, through participant-observation, to learn about the workings of the society or culture.

Rappaport not only took on the job of describing and understanding the inner workings of Tsembaga culture, he also attempted to understand this in the context of Tsembaga ecology. In the second case, Thomas's task was no less daunting, but his work was done within the framework of an integrated project. This was the Andean Biocultural Studies project, initiated by Paul Baker (Baker and Little, 1976) at the Pennsylvania State University in the U.S., with the primary objective of studying the patterns of adaptation of high-altitude natives to the hypoxia and cold of the Andean altiplano. When Thomas began his work, several years of data on social conditions, nutrition, human physiology, demography, and weather conditions had already been collected, and the area had been mapped.

Rappaport's Work

Rappaport's research was reported in a now famous book entitled Pigs for the Ancestors, which was published in 1968, and reprinted in 1984 with a 190-page Epilogue, in which he addressed his critics and reevaluated some of the research. The work is brilliant, in that it addresses some of the fundamental issues underlying anthropological theory, including: social control, environmental causality for behaviour, and the connection between individual behaviour and cultural norms or prescribed social behaviour.

In the work, Rappaport suggested first, that human population numbers, pig population numbers, the warfare cycle, agricultural productivity, patterns of exchange of goods, the distribution of land and people, and the maintenance of the ecosystem as a productive system were all tightly interrelated as a working system. Second, he suggested that the system was in a state of equilibrium maintained by feedback mechanisms. And third, and perhaps most controversial, that the regulating or controlling mechanism that kept the system going was the information provided in the form of ritual and a ritual cycle. Within this research, he took both a materialistic and a functionalist approach to social science, he identified human behaviour as adaptive in the context of the social and ecological systems, and he identified human behaviors as subject to selection of favorable behaviors in the context of maintenance of the human/ecological system.

Needless to say, and despite the fact that ecological anthropology was in vogue at that time, Rappaport's critics in sociocultural anthropology were severe in their verbal assaults. His work had attacked some of the fundamental icons of anthropology. He and other ecological anthropologists were accused of:

(1) reifying the ecosystem (to treat the abstraction of an ecosystem as if it had material existence);
(2) vulgar materialism (a belief that the materialistic approaches used in ecological anthropology were simplistic in their social context);
(3) a calorific obsession (placing too much emphasis on flows of energy through the system);
(4) excluding historical factors (too much emphasis on equilibrium and stability in diachronic state in the systems studied);
(5) setting up false boundaries (human cultures go beyond ecosystem boundaries);
(6) shifting levels of analysis (applying one level of interpretation to another); and
(7) dealing with an "impoverished" ecosystem approach (in contrast to "evolutionary ecology").

Scientists in the ecology community were debating some of these issues, but in many cases, anthropologists did not fully understand the bases for the debates, plus the human dimension added profound levels of complexity to these issues. In any case, the emotion behind these critical writings and the use of such intemperate terms such as "reification," "vulgar," "obsession," "false," and "impoverished" reflected the intense feelings about ecological and cultural materialistic approaches by a majority of anthropologists.

Thomas's Work

The work that Brooke Thomas conducted on energy flow research in a highland native community in Peru was begun in the late 1960s after Rappaport's and Vayda's ecological studies of New Guinea populations. The work was stimulated by Rappaport's and others' research and by the ecologist H.T. Odum's graphic shorthand language to represent the flows and controls of energy through ecosystems. At the time of the study, Quechua natives of the altiplano employed a mixed subsistence of cultivation of potatoes (and other tubers) and quinoa (chenopods) and herding of llamas, alpacas, and sheep.

By comparing food energy production (outputs) with labour expenditures (inputs), Thomas demonstrated that cultivation provided a 10: 1 return, while livestock herding provided only a 2: 1 energy return (Thomas, 1976). Animal products (meat, hides, wool) were highly prized at lower elevations; hence, trade of animal products for other foods (e.g., maize, sugar) increased the ratio to more than 7: 1. Thomas's model, although representing averages and a simplified view of the energetics of production and expenditure in this community, nevertheless quantitatively demonstrated the utility of some of the principles of Quechua native subsistence through energy flow.

Thomas's work was the focus of an intense critique in a book called Energy and Effort that was edited by the distinguished human biologist Geoffrey A. Harrison (1982). The critique was penned by Philip Burnham, who began his comments by criticizing H.T. Odum's (1971) work on Environment, Power, and Society, identifying it as "reductionist" and "breathtakingly naive."

This book was somewhat naive, particularly in its chapters on human politics and religion, but many of the analytical approaches were very useful. Burnham (1982) continued his comments by outlining methodological problems that he saw as limiting understanding of human behaviour by energy flow studies. One point has merit, where he stated: "...there is the problem of the multidisciplinary competencies required of a single researcher engaged in human ecological field study..." (Burnham, 1982: 233). Other arguments that he made were:

(1) that the costs are too high for the "pay-offs" of energetics (anthropologists have grown accustomed to very modest research budgets);

(2) adequate nutritional assessment is impossible from field studies (Michael Latham, an eminent nutritional scientist from Cornell University once told me that it was really the anthropologist who could address several key nutritional issues from extended field work);

(3) too many simplifying assumptions were made (this is a key to modeling, but only at the outset);

(4) it is impossible to account for all of the social issues (but, this is never even possible in sociocultural analyses);

(5) it is inappropriate to apply the functional/adaptational paradigm borrowed from biology (this reflects the hostile views toward the biological sciences that many social scientists feel). Adaptation as a concept was criticized heavily where he expressed his view on "...the inadequacy of the concept of adaptation as applied to social behaviour!" In brief, Burnham typified the views of many sociocultural anthropologists (despite his interests in human ecology) where a materialistic, adaptational, quantitative approach that draws on basic biological principles somehow sidesteps the fundamental bases of human culture and society.

ECOLOGY

Ecology is the study of the "relationships among living organisms or between them and the physical environment."

Some characteristics of the scientific study of ecology:

- Macro-scale ecologists (global-scale ecology, '60's) have lost political ground to micro-scale ecologists.
- Landscape ecology recognizes human influence on non-human species, yet persist in distinguishing human from natural landscapes.
- Human ecology concentrates on systematic and evolutionary aspects, while social ecology emphasizes behaviour. Both study human-environmental relationships in distant past or present. Sociobiology is a paradigm within human social ecology
- Cultural ecology and cultural geography examine adaptive strategies. Both are cognizant of the role of culture in human adaptation, but not interested in long term change.
- Environmental History is the intellectual history of the environmental movement, including the political and economical implications of environmental interaction.
- Environmental Ethics explores value systems as they relate to human conduct.

None of these fields has truly integrative approach, and many lack an explicit historical component. What is needed is a multi-scalar temporal and spatial frame with an explicit focus on the role of human cognition in the human-environmental dialectic.

Historical Analogs

Global climate change is one of the most pressing event of current times. The anticipated changes demand investigations into patterns of human adaptation to climatic variability and change.

However, the global climate change models used by physical scientists to predict climatic changes do not discriminate among biotic zones or anywhere near a human scale. Furthermore, many physical scientists assume that "novel circumstances" render any historical analogy to current anticipated global climatic change irrelevant. This attitude is due to:

(1) The lack of high quality long term (>100 yr.) instrumentally obtained data
(2) Local proxy data (such as tree ring) are only valid at the broadest temporal scales.
(3) Dismay of the comparative messiness of soft social science data
(4) Vested interest in favour of novel technologies and undervalue of traditional solutions

A regional approach could overcomes this. A region's air mass data, hydrology, soil, topography and species distribution can be used in regional models. Regionally documented ethnography, archeology, and documentary evidence evidences results of human activities and past choices which encompass the entire system.

Multiple regional environmental changes can identify sensitive geographical locations. Interregional relationships may then be established and integrated with global data. This approach fosters creativity and the development of local and regional answers to global situations in which sensitive cultural issues play an important part.

MODELLING CLIMATE CHANGE

Climate models attempt to simulate the behaviour of the climate, in an attempt to understand the key physical, chemical and biological processes which govern climate. Climate models give us a better understanding of the climate system, providing us with a clearer picture of past climates by comparison with records of instrumental and palaeoclimatic observations, and enabling us to predict future climate change.

Models can be used to simulate climate on a variety of geographical scales and over different periods of time. The basic laws and other relationships necessary to model the climate are expressed as a series of mathematical equations. The climate however, is a very complex system, and supercomputers are needed for the task. Global climate models have been used extensively to project global warming in the 21st century due to mankind's greenhouse gas pollution of the atmosphere.

Estimates of future increases in greenhouse gases are inputted into the model, which then calculates how the global climate might evolve or respond in the future to the enhanced greenhouse effect.

Although climate models can aid understanding in the processes which govern the climate, the confidence placed in such models should always be questioned. Critically, it should be remembered that all climate models represent a simplification of the climate system, a system that may ultimately prove to be too complex to model accurately.

Climate models must therefore be used with care and their results interpreted with due caution. Margins of uncertainty should be attached to any model projection. Results from climate models should always be validated or tested against real-world data, including both instrumental and palaeo-climatic records where available.

PALAEOCLIMATE CHANGE

The global climate has shifted and varied for billions of year, perhaps since the Earth first had an atmosphere. The oldest palaeoclimatic records have allowed us to reconstruct climate fairly reliably during the last 500 million years. Over this time, the global climate has moved from extensive periods of global warmth to periods of global cold several times, each lasting 100 million years or more. Although today we are concerned about global warming, we do in fact lie in the middle of global ice-house climate, which began 40 million years ago, when the first permanent ice sheets formed on Antarctica. The change from the much warmer global climate which existed during the age of the dinosaurs, when global average temperature was perhaps 10°C higher than at present, is thought to have been caused by changes in the distribution of landmasses and the associated changes to energy redistribution throughout the climate system.

Within the long-term global icehouse climate, much shorter-term fluctuations in global climate have occurred. Relatively cold periods known as Ice Ages or glacials, each lasting roughly 100,000 years, are interspersed with much shorter warmer episodes or interglacials, lasting only 10,000 years. We now have a relatively clear record of such climatic fluctuations over the last 2 million years. Currently, the global climate lies within an interglacial. Global average temperature 20,000 years ago towards the end of the last Ice Age was some 5°C lower than today, when the north polar ice sheets were expanded to cover a considerably greater area of the continental Northern Hemisphere than is the case today. These glacial-interglacial fluctuations are believed to be driven by changes in the position of the Earth in its orbit around the Sun, and enhanced by climatic feedback processes which involve changes in ocean circulation and the greenhouse gas composition of the atmosphere.

Within the latest present interglacial period, further fluctuations in the global climate can be seen in the palaeoclimatic records and more recently in the instrumental records. During the last 1000 years, the climate has moved from a period of Medieval warmth to a"Little Ice Age" between the 16th and 19th centuries, with changes of between 0.5 and 1°C in the global average surface temperature. Although it is not clear what has caused these climatic changes, variations in the Sun's energy output, ocean circulation and the occurrence of major volcanic eruptions are believed to play a part.

Most recently, we have entered a renewed period of global warming since the beginning of the 20th century that we suspect is the result of mankind's enhancement of the natural greenhouse effect through the pollution of the atmosphere. Global average temperature is now about the same as it was during the Medieval warm period, although still much lower than it was 100 million years ago during the age of the dinosaurs. Climate change that predates the instrumental period of direct weather observations is known as palaeoclimate change. Palaeoclimatology provides a longer perspective on climate variability that can improve our understanding of the climate system, and help us to predict future climatic changes as a result of man-made global warming.

Evidence for palaeoclimatic change can be obtained by the study of natural phenomena which are climate-dependent. Such evidence comes from palaeoclimatic records.

Many natural systems are dependent on climate. From these it may be possible to derive palaeoclimatic information for variables such as temperature or rainfall. Such proxy or indirect records of climate contain a climatic signal. Deciphering that signal is often a complex business.

Four types of palaeoclimatic record are commonly analysed for past climate variations. These include historical records, glaciological records, biological records and geological records.

The suitability of a particular palaeoclimatic record for reconstructing a past climate will be largely dependent upon the time scale of past climate change under study. Recent climate changes during the last few thousand years can be reliably reconstructed from tree ring analyses, which often yield continuous records and provide high-resolution (annual or even seasonal) data.

Over the longer term, ice cores and sea sediments offer information about palaeoclimates stretching back hundreds of thousands of years, although the data resolution may not be as fine.

Generally, the further back in time we go the greater the margins of uncertainty that will be attached to palaeoclimatic reconstructions.

Palaeoclimatology

Palaeoclimatology, from the Greek word"palaios", meaning"ancient", is the study of past climates and past climate change, prior the period of instrumental records. The study of palaeoclimatology can encompass much of Earth Antiquity, or at least that part of it for which reliable palaeoclimatic records are available to reconstruct palaeoclimates.

Palaeoclimatology may be distinguished from climatology and contemporary climate change, which studies present day climate and climate changes restricted to the most recent period (the last 150 years) since instrumental records of daily weather observations have become available.

To reconstruct palaeoclimates, palaeoclimatologists cannot use direct observations of temperature, rainfall and other climatic variables. Instead they use proxy records of natural phenomena which are climate-dependent. These include analyses of tree rings, ice cores, sea sediments and even rock strata exposed at the Earth's surface which may hold clues to the state of the climate millions of years ago.

Climate models run on computers may also be used to test theories about possible mechanisms of palaeoclimate change.

Sea Sediments

Billions of tonnes of sediment accumulate in the ocean basins every year. The nature of such sediments may be indicative of climate conditions near the ocean surface or on the adjacent continents. Sediments are composed of both organic and inorganic materials.

The organic component of sea sediment includes the remnants of sea-dwelling microscopic plankton, which provide a record of past climate and oceanic circulation. For example, by studying the chemical composition of plankton shells, we can reveal information about past seawater temperatures, salinity (saltiness), and nutrient availability. Indeed, such techniques have been used to reconstruct ocean temperatures over the last 100 million years, and have confirmed continental drift theories of climate change that a long term global cooling has taken place since the extinction of the dinosaurs.

Most inorganic material comes from adjacent landmasses, eroded from rocks and washed down to the coast by river channels, or blown from soils, dusty plains and deserts. The nature and abundance of inorganic materials provides information about how wet or dry the nearby continents were, and the strengths and directions of winds.

Sun

For many years scientists have speculated that changes in the amount of energy given off by the Sun can influence the Earth's climate. There is no doubt that variations do occur in various characteristics of the Sun on a range of time scales. The 11-year cycle in the number of sunspots on the face of the Sun is well known. But other parameters, including the size of the Sun, vary too, and over different time scales from tens of years to thousands of years.

What is less clear is whether or not these changes produce significant variations in the total energy output of the Sun. The total solar energy received by the Earth, or solar constant, has only been measured accurately since the advent of the satellite era, 40 years ago. In addition, changes which have been detected over the past 20 years are very small in magnitude, and probably too small to account for all of the observed global warming that has occurred during this period. While changes in total solar energy may be greater on longer time scales, this is only a speculative possibility. Nevertheless, scientists have proposed that longer-term changes to the nature of the sunspot cycle may have been the cause of the Little Ice Age that occurred between the 16th and 19th centuries.

CLIMATE CHANGE ON ECOSYSTEMS

One way that scientists gain understanding of how global warming will affect ecosystems is to analyse the effects of past climate on paleoecosystems. A paleoecosystem is an ecosystem that existed in a former geologic time period. By relating vegetative cover to past climates, models can be developed. Once a reliable model is created, input variables such as CO_2 can be varied and the results analysed. An example of what the effect would be like on populations of Douglas fir in the northwestern United States if the CO_2 content in the atmosphere were double what it was before the Industrial Revolution was modeled by the U.S. Geological Survey and is shown in the illustration that follows. In a study conducted by the U.S. Global Change Research Programme in Washington, D.C., first in 2001, then updated in

2004, in trying to predict what the effects of future climate change would have on ecosystems, they concluded that"climate change has the potential to affect the structure, function, and regional distribution of ecosystems, and thereby affect the goods and services they provide." They based their conclusions on a modeling and analysis project they conducted called the Vegetation/Ecosystem Modeling and Analysis Project.

This project was used to generate future ecosystem scenarios for the conterminous United States based on model-simulated responses to both the Canadian and Hadley scenarios of climate change. The VEMAP was subsequently used in a validation exercise for a Dynamic Global Vegetation Model by Oregon State University and the U.S. Forest Service in 2008. Their MC1-DGVM was used as the input data in both the VEMAP and VINCERA. Their MC1 was run on both the VEMAP and VINCERA climate and soil input data to document how a change in the inputs can affect model outcome. The simulation results under the two sets of future climate scenarios were compared to see how different inputs can affect vegetation distribution and carbon budget projections.

The results indicated that"under all future scenarios, the interior west of the United States becomes woodier as warmer temperatures and available moisture allow trees to get established in grasslands areas. Concurrently, warmer and drier weather causes the eastern deciduous and mixed forests to shift to a more open canopy woodland or savannatype while boreal forests disappear almost entirely from the Great Lakes area by the end of the 21st century. While under VEMAP scenarios the model simulated large increases in carbon storage in a future woodier west, the drier VINCERA scenarios accounted for large carbon losses in the east and only moderate gains in the west. But under all future climate scenarios, the total area burned by wildfires increased." The similarities of the two models served to validate the VEMAP project.

The Hadley model was developed by the Hadley Centre for Climate Prediction and Research in England. Also referred to as the Met Office Hadley Centre for Climate Change, it is based at the headquarters of the Met Office in Exeter. It is the key institution in the United Kingdom for climate research. It is currently involved not only with understanding the physical, chemical, and biological processes within the climate system, but also with developing working models to explain current phenomena and to predict future climate change. It also monitors global and national climate variability and change and strives to determine the causes of the fluctuations. The Canadian climate model was developed by the Canadian Centre for Climate Modelling and Analysis. The CCCma is a division of the Climate Research Branch of Environment Canada based out of the University of Victoria, Victoria, British Columbia.

Its specific focus is on climate change and modeling. In the past nine years, the CCCma has produced three atmospheric and three atmospheric/oceanic general circulation models, making them one of the international leaders in climate change research.

What they found was that over the next few decades climate change in the United States will most likely lead to increased plant productivity as a result of increasing levels of CO_2 in the atmosphere. There will also be an increase in terrestrial carbon storage for many parts of the country, especially the areas that become warmer and wetter. The southeast will most likely see reduced productivity and, therefore, a decrease in carbon storage. By the end of the 21st century, many areas of the country will have experienced changes in the distribution of vegetation. Wetter areas will see the growth of more trees; drier areas will have drier soils that will cause forested areas to die off and be replaced by savanna/grassland ecosystems.

Modeling the vegetation evolution and adaptation is more difficult. The study focused on two time periods: 2025-2034 and 2090-2099. In the near term, biogeochemical changes are expected to dominate the ecological responses.

Biogeochemical responses include changes based on the natural cycles of carbon, nutrients and water. The responses are affected by changing environmental conditions such as temperature, precipitation, solar radiation, soil texture, and atmospheric CO_2.

It is these natural cycles that affect carbon capture by plants with photosynthesis, soil nitrogen processes, and water transfer. These biogeochemical factors are what influence the production of vegetation. In the results from the near-term biogeochemical model, the scientists concluded that there would be an increase in CO_2. They estimate that currently the average carbon storage rate is 66/Tg/yr. The Hadley model predicted that carbon storage rates by 2025-2034 would increase to 117/Tg/yr. The Canadian model estimated CO_2 to increase 96 Tg/yr.

The Canadian model projected that the southeastern ecosystems will lose carbon in the near term because they predict the climate there will become hot and dry. The biogeography models look at the changing landscape based on changes in CO2, evapotranspiration, vegetation establishment, and competition between species, growth rates, and life cycle/mortality rates. In this model, scientists at both the Hadley and Canadian Centre for Climate Change agree that vegetation will be able to freely move from one location to another.

Changes in vegetation distribution will vary from region to region as follows:

- *Northeast*: Forests remain the dominant natural vegetation, but forest mixes will change. There will also be some increase in savannas and wetlands.

- *Southeast*: Forests remain the dominant ecosystem, but mixes change. Savannas and grasslands encroach on forests, especially towards the end of the 21st century. Drought and wildfires contribute significantly to forest destruction.
- *Midwest*: Forests remain the dominant land cover, but changes in species type occur. There will be a modest expansion of savannas and grasslands.
- *Great Plains*: Slight increase in woody vegetation.
- *West*: The areas of desert ecosystems shrink, and forest ecosystems grow.
- *Northwest*: Forested areas grow slightly.

A separate study conducted by the Canadian government predicts the following ecosystem changes as a result of changing climate over the next 100 years:

- *Coastlines*:
  - Flooding and erosion in coastal regions
  - Sea-level rise
- *Forests*:
  - Increase in pests
  - Increased levels of drought and wildfire
- *Plants and animals*:
  - Warmer temperatures could make water supplies more scarce, having a negative impact on plants and animals, not giving them time to adjust.
- *Crops*:
  - In some areas, warmer climate may allow a three- to five-week extension of the frost-free season, which could benefit commercial agriculture.
  - In other areas, drier soils and lack of water will have a negative impact on agricultural productivity.
- *Wells*:
  - The quality and quantity of drinking water may be threatened by increasing drought.
- Harsh weather:
  - Winter storms, floods, drought, heat waves, and tornadoes could become more frequent and severe.
- *Fisheries*:
  - Populations and ranges of species sensitive to changes in water temperature will be negatively affected.
  - Salmon harvests will be lower in the Pacific.

  - Changes in ocean currents may have a negative impact on the fisheries in the Atlantic.
- *Lakes and rivers*:
  - Water levels will decline under the influence of drought, negatively affecting drinking water quality.
  - Use of lakes for transportation, recreation, and fishing, and the ability to generate electricity may be curtailed under droughtlike conditions.
  - Other areas that may have an increase in precipitation may experience flooding, rising sea levels, and severe storms.

In February 2005, the Met Office in Exeter, England, issued a report titled"Avoiding Dangerous Climate Change." The objective of the study was to determine what levels of CO_2 were considered the tipping point for dangerous climate change with harmful effects on ecosystems and what actions could be taken now to avoid such an outcome. In the report, then prime minister Tony Blair stated:"It is now plain that the emission of greenhouse gases... is causing global warming at a rate that is unsustainable." Environment Secretary Margaret Beckett stated,"The report's conclusions would be a shock to many people.The thing that is perhaps not so familiar to members of the public... is this notion that we could come to a tipping point where change could be irreversible. We're not talking about it happening over five minutes, of course, maybe over a thousand years, but it's the irreversibility that I think brings it home to people." The report, published by the British government, says there is only a small chance of greenhouse gas emissions being kept below"dangerous" levels. It warns that the Greenland ice sheet could melt, causing sea levels to rise by 23 feet over the next 1,000 years.

It also warns that developing countries will be the hardest hit. The report also states, based on the vulnerability of many of the world's ecosystems, that the European Union (EU) has adopted a target of preventing an increase in global temperature of more than 3.3°F (2°C). Some believe even that may be too high.

The report states:"Above two degrees, the risks increase very substantially," with"potentially large numbers of extinctions" and"major increases in hunger and water shortage risks... particularly in developing countries." In order to meet their goals, British scientists have advised that CO_2 levels should be stabilized at 450 parts per million (ppm) or below. Currently the atmosphere contains 380 ppm. In response to this, the British government's chief scientific adviser, Sir David King, said that was unlikely to happen. He stated,"We're going to be at 400 ppm in 10 years' time. I predict that without any delight in saying it." Myles Allen, an expert on atmospheric physics at Oxford University, said that:"Assessing a 'safe level' of carbon dioxide in the atmosphere was 'a bit like asking a doctor what's a safe number of cigarettes to smoke per day.'"

The report does conclude, however, that there are technological options available to reduce CO_2 emissions that will need to be used. The study also concluded that the biggest obstacles involved with using these new technologies, along with renewable resources of energy and"clean coal," are the current economic investments and traditionally strong bond to the oil industry, cultural attitudes that oppose change, and simple lack of awareness by many people. Various conservation organizations currently involved in the battle against global warming, such as the Union of Concerned Scientists (UCS), the Defenders of Wildlife, and the World Wildlife Fund (WWF), also support these ideas.

CONCERN FOR BIODIVERSITY

An indication of society's concern for biodiversity can be gleaned from opinion surveys, even though these have only been undertaken recently. An analysis of national surveys from 1975 to 1994 identified high levels of concern over environmental issues, and biodiversity was consistently important within the broader environmental field. During the 1990s, the ABS undertook more regular surveys (ABS 1999a). In 1999, the environment was nominated as the most important social issue by 9% of the Bureau's sample, above issues such as crime and health. Of those who did not rank the environment as their top issue, 69% stated that they were concerned about environmental issues. This represents a slight decrease on recent years although in 1986 the corresponding figure was 49%. The issues of concern most directly relevant to biodiversity, destruction of trees/ecosystems and of animals/wildlife, were identified by 29% of people, and 43% believed that the quality of the environment had decreased in the 1990s.

DOMESTIC LAW AND BIODIVERSITY

One indication of change is the expression of biodiversity in domestic law. Integrated nature conservation legislation, which caters for harvesting control, conservation reserves and species protection, and specific legislation, which deals with threatened species, only date back a few decades in Australia (most states enacted such laws in the 1970s). By 1999, over 120 state, territory and Commonwealth statutes expressly referred to ESD as an objective and set of guiding principles. There has also been a growth in the number and scope of international instruments concerning biodiversity.

TYPES OF BIODIVERSITY

Biodiversity is a generic term that can be related to many environments and species, for example, forests, freshwater, marine and temperate environments, the soil, crop plants, domestic animals, wild species and micro-organisms. Basically it can be classified according to three types of diversity:

- ecosystems and landscapes (habitat diversity)
- animal, plant, bacterial species (species diversity)
- all genes (genetic diversity)

Of particular importance are the taxonomically isolated species, as they have little similarity to other species and therefore are unique with respect to their genetic structure. These species are often endemic meaning limited to one specific area. Their extinction would mean a greater loss for global biodiversity rather than just the extinction of a species.

Distribution of Biodiversity

In Europe, biodiversity is very unevenly distributed with the least variety of ecosystems and lowest diversity occurring in Northern Europe. Centres of high biodiversity are to be found in the Mediterranean (Italy, Spain, Greece, France) and on the fringes of Europe (Bulgaria, Ukraine, Georgia, Armenia, Turkey), with over 5,000 endemic plant species that occur only in these countries. The Mediterranean is Europe's richest sea in terms of biodiversity.

The Importance of Biodiversity

Biodiversity is often used to draw attention to issues related to the environment. It can be closely related to

- the health of ecosystems. For example, the loss of just one species can have different effects ranging from the disappearance of the species to complete collapse of the ecosystem itself. This is due to every species having a certain role within an ecosystem and being interlinked with other species.
- the health of mankind. Experiencing nature is of great importance to humans and teaches us different values. It is good to take a walk in the forest, to smell flowers and breath fresh air. More specifically, natural food and medicine can be linked to biodiversity.

Biodiversity in the Black Sea

The Black Sea region used to be one of the most important areas for fisheries and for food and income for local people. Sturgeons (Acipenser sp.), mullets (Mugil sp.), and mackerel (Scomber sp.) as well as other species have been extensively exploited in this area.

However, human activities related to agriculture, shipping and tourism now exert strong pressure on the environment, especially in the northern part of the Black Sea, and are taking their toll on biodiversity and damaging fish stocks.

Threats to Biodiversity

Large amounts of fertilizers were carried to the sea from agricultural areas along the large rivers, such as Don, Dnepr, and Dnjestr. Oil leaked into the sea from ports on the eastern shores and directly from tankers crossing the Mediterranean. Wastewater from cities and heavily visited coastal tourist towns was discharged without any purification.

Consequences for the Ecosystem

All of this leads to pollution and eutrophication meaning the enhanced blooming of planktonic algae due to increased nutrient loads. Planktonic algal blooms lower the water transparency letting only a little amount of light through the water where makrophytes usually grow. This is how the natural belt of bottom vegetation along the Black Sea coast has been destroyed. For example, the vertical distribution range of Cystoseira spp. decreased from 0-10m to 0-2,5m.

The consequences were drastic because the habitat is of major importance as a nursery for spawn and hatchlings of many marine species. This led to a drop in reproduction rates and hence fish stocks.

Biodiversity of the Coastal Zone

Coastal areas are exceptionally productive environments, rich in natural resources, biological diversity and with a high potential for commercial activity. The importance of biodiversity in the coastal zone can be demonstrated by 8 out of the 40 EU listed priority habitats of wild fauna and flora falling into the coastal habitat. Approximately a third of the EU's wetlands are located on the coast as well as more than 30% of the Special Protected Areas designated under the Directive for the conservation of wild birds. The reproduction and nursery grounds of most fish and shellfish species of economic value also comes from this area, which accounts for almost half of the jobs in the fisheries sector.

Pressure on Coastal Biodiversity

Coastal areas are increasingly vulnerable to stresses from both human activities and the forces of nature. The complexity of human activities, natural systems and ownership in the coastal zone, requires an integrated management scheme to allocate coastal resources efficiently and minimize environmental degradation. Choices have to be made between competing uses and limits of resource exploitation if escalating conflicts and resource degradation are to be avoided.

The attitudes of community and industry to the use of biological resources should change from the 'maximum yield' approach to one of 'ecologically sustainable', which recognizes the need for conservation of biological diversity and maintenance of ecological integrity. Integration of management regimes within and between different sectors to meet environmental, economic and social objectives must be realized in order to achieve sustainable development.

Integrated Approach

Integrated policies will also provide the opportunity for all the people to accept responsibility for their actions and the impact they may have on biological diversity. The development of integrated policies for managing the coastal resources is necessary:

- to coordinate activities within and between all levels of government;
- to ensure that full social and environmental consequences (and costs) of development activities are considered;
- to ensure that the public interest is properly taken into account.

ASSIGNING ECONOMIC VALUE TO BIODIVERSITY

Biodiversity refers to the diversity of life in all its forms and all its level of organisation, not just plant, animal and microorganism species. At its most elemental level, biodiversity, encompasses the varied assemblages of organic molecules that comprise the genetic basis of life. On the other end of the spectrum there are biomes-the vast stretches of tundra, desert, forest, ocean, etc. In between come-population, race, sub-species, community, ecosystem.

Why is Biodiversity Important?

For many people the very questioning of the worth of biodiversity is illicit. They would argue that humankind has a moral obligation to conserve biodiversity, an obligation that comes with the fact that humans have the capability to destroy much of that biodiversity. Others find a religious support for such a view there is some stewardship responsibility on behalf of some deity. One problem with these moral views is that they often conflict with other moral views about, say, the right to earn a living, the right to have access to basic needs such as food, cloth and shelter, and so on.

If conserving biodiversity conflicts with those rights, then some 'meta-ethical' principle is required for deciding which moral view should prevail. One aspect of the process of changing popular perceptions about biodiversity resources is to show that the sustainable use of biodiversity has positive economic value and that this economic value will often be higher than the alternative resource uses which threatens biodiversity.

Valuing Biological Diversity Conservation

The issue at hand concerns the measurement of natural capital, great difficulties arise in assigning values to natural capital precisely because the market in which it is traded are limited. In the past few decades a number of techniques have been developed to place monetary value to the benefits of conserving an area for biodiversity. Some rely on market prices of related goods and services both to value benefits and to estimate costs, while other rely on survey based approaches to infer values.

Hedonic Pricing Method

Property values are affected by a number of variables including environmental quality. After excluding other variables, including environmental quality, the residual price difference can then be ascribed, at least theoretically, to differences in environmental quality, e.g., the increased value of property located next to natural areas. According to a report by Nelson (1982) traffic noise, measured in Leq (equivalent continuous sound level), a one unit change produces property price depreciation of 0.5-1.0%.

Ravel-cost Method

This approach looks at the pattern of recreation use of a natural parks and uses this information to derive a demand curve to estimate the total amount of consumer's surplus. For example, travel behaviour reveals that Costa Rican visitors are willing to pay USD 35 per household to visit a tropical rainforest site in Costa Rica. Foreign visitation is likely to be worth far more than domestic, as foreign visitors have higher travel costs.

Contingent Valuation Method

The Contingent Valuation Method (CVM) bypasses the need to refer to market prices by asking individuals explicitly to place values upon environmental assets. This is also referred to as an expressed preference method. An interesting advantage of the CVM approach is that it can, in theory, be used to evaluate resources, that people have never visited personally e.g. the Antarctica which people are " Willing to Pay" to preserve but would not in general ever want to visit.

Consumptive Benefit Method

There are a number of products which can be harvested on sustainable basis and marketed. For example, a probable number of higher plant species, which are widely used as basis for pharmaceutical drugs, is some 500,000. According to Pearce and Moran (1994) economic value of these plants can be approached by looking at:

(i) the actual market value of the plants when traded,
(ii) the market value of the drugs of which they are the source materials,
(iii) the value of the drugs in terms of their life-saving properties, and using a value of a 'statistical life',
(iv) the lost pharmaceutical value from disappearing species.

As there is absence of 'global markets' in the benefits of biodiversity, the developing countries face major problems of appropriating the global benefits of sustainable use of biodiversity. As long as these global values cannot be captured by host countries, biodiversity will be a risky investment in many contexts.

THE VALUE OF BIODIVERSITY-THREATENED HABITATS

Biodiversity is not just some benign backdrop for hiking holidays, but the very substance and foundation of our survival, whether we realize it or not. We are entirely dependent upon the plants, animals, fungi, and micro-organisms that share the world with us. Individually, they alone feed us, and without them we would starve. Yet we frequently act to undermine these very species essential to our welfare. In addition to food, they provide many of the drugs and other medicinal and industrial products on which the quality of our lives increasingly depends. They offer the promise of sustainable economy-productivity that the Earth can support on a continuing basis, so our children and, in turn, their children will survive and be able to live peaceful lives of abundant splendour.. We live in an age driven by the apparently insatiable desire of industrialized nations to go on getting richer, and of free-market economics driven by equally insatiable multi-national corporate organizations, competing to exploit the remaining resources of the planet. Though we are now beginning to consider atmospheric changes such as the ozone hole and global warming as significant international problems, we have yet to demonstrate we can hold good to effective action.

However at the same time the threats to biodiversity, which are much more long-lasting are not being taken seriously politically. The fact is that up to a quarter of the species on Earth may be lost in the course of the next three decades-within the lives of the majority of us alive today, and a majority of biodiversity is likely to have perished by the middle of next century.

"Each year, we are cutting and burning 1.5 to 2 per cent of the world's remaining tropical rainforests; losing an estimated 24,000 million tonnes of topsoil; and adding some 93 million people to a world that is already far too full, judging from the extent of human misery, and starvation, let alone from the depletion of every conceivable resource." (Peter Raven "Porritt 71"). Since Rio, the rate of felling of the Amazon has increased some 30%, finally exacerbated by drying from accentuated El Nino, a possible consequence of global warming.

Every, point on the Earth's surface, from the frozen wastes of Antarctica to the most remote stretches of the oceans, receives a steady shower of man-made chemicals. We are clearly "managing" the entire planet now, for better or for worse, whether we acknowledge this responsibility or not. In this sense, there is no longer any region that can be said to be truly "natural". The future impact of genetic engineering technologies is likely to be vast and has barely even been considered in the midst of runaway technological initiative.

Stewardship of Planet Earth

"Will we act as responsible stewards of the many organisms that share the Earth with us? We have certainly not given much evidence so far of our commitment; having given names to only 1.4 million of them, we don't know whether the total number may be 10 or 100 million. We understand even less-often nothing at all-about their individual properties, or the ways in which they interact with one another. Moving beyond our growth-orientated mentality which assumes that every, productive system on the planet can be expanded indefinitely to meet our needs, regardless of its biological basis, is an essential ingredient of stewardship. Every nation must work to develop its own base of information on biodiversity, and strive to understand, to use, and to save it, both for its own purposes and for future generations. For rich nations, this means understanding that we cannot continue to ravage our strictly limited home planet as if its productivity and stability were simply inexhaustible. For poorer nations, the challenge will be even greater, and will not be met without reversing the tragic flow of many millions of dollars from poor, starving countries to the rich industrialized North. Environmental stewardship and social justice go hand in hand".

Multifarious Insects

There are more insects in the world than any other group of organisms. Until recently, virtually all the natural medicines we have developed have come from plants and fungi, but there is no reason why a sector which produces both cochineel and potent stinging toxins cannot have diverse biochemicals, especially to resist predators and to give protection against diverse plant toxins. "Currently only about 20% of all insect species have been identified, let alone chemically characterized.

Bioprospecting in the insect species has only begun and is barely explored, yet the diverse insect species could be rapidly reduced in number with the destruction of habitats" (Peter Raven "Porritt 71"). There are currently some efforts under way to develop cooperative relationships with the governments of some developing countries in which wide assays of insect species are carried out.

A Case: Cone Snails

Just as sarrow poison frogs are famed for their toxins, so are cone snails. Typical cone snails contain between 200 (chemical analysis) and 2000 (genetic analysis) types of polypeptide toxin of between 10 and 30 amino acid units. By contrast, snake toxins frequently contain 80 peptides and spider toxins up to 1000 units. Magic sponge Sea creatures could be a goldmine for powerful drugs

Left: Detail of the shell pattern of Olivia porphyria and wave-like cellular automaton with long-range and local autocatalytic interactions. The cone snail toxins represent both acetylcholine blockers similar to cobra venom toxin and ion blockers, which instead of the stopping flow of sodium ions like puffer fish, block one of the seven types of calcium channel. They also contain glutamate NMDA receptor blockers. No two species have the same toxins and many change their output over time. Their extremely varied shells, which mimic know cellular automata have fetched more than a Vermeer painting.

Scientific American Feb 96 reported their use in a new pain-killing drug SNX-111 which proved effective in 5 out of 7 cases of intractable pain which had become tolerant to opiates. It thus represents the first of an entirely new class of compounds to treat severe pain. A second case of medicinal toxins: The Platypus New Scientist Jan 3 98. The poison from the spurs of the duck-billed platypus could point to ways of designing new types of painkiller, physiologists say. They believe that a toxin in the poison acts directly on our pain receptors. The male of this strange Australian species has spurs on its hind legs that contain at least four different toxins. They are thought to help the animals defend their breeding territory. "It's an unusual venom that's not designed to kill or paralyse its victims," says Rosemary Martin of the Australian National University in Canberra.

Humans that are pricked by the platypus experience intense pain that can sometimes last for several weeks, she adds. "it seems designed as a deterrent to induce pain." Martin and her colleagues, led by Greg de Plater, have now pinned down how at least one component of the poison of the platypus's spurs behaves. The researchers tested the effects of a protein in the toxin on the neurons of laboratory mice. It turned out that the protein binds to a channel in the membranes of the neurons which allows positive ions to enter and leave the cells.

Snake oil and Cancer The protein contortrostatin named after the copperhead has been found to have a dual anti-cancer action stopping cells spreading and reducing blood-vessel formation. The African saw-scaled viper also produces a heart drug Aggrastatis.

Case Study: The Wollemia Pines

In a remote stretch of national park in Australia, was accidentally discovered this stand of an exceedingly ancient pine Wollemia nobilis known only from fossil records dating back to 250 million years. The trees stand in a formation of Wollemi sandstone dating itself from around 250 million years ago. Wollemia was common in Pangea from 200 million years and existed worldwide until 65 million years ago and continued in Gondwanaland up until 30 million years ago (New Scientist 1997). This single stand later proved to be genetically monoclonal suggesting it has survived by vegetative sprouting from the forest floor. A second stand has been found a kilometre away. Otherwise this is the only specimen of this ancient and widespread species still on Earth.

Endemic Diversity: Hot Spots and Fragile Niches

Almost every corner of the earth supports some form of life but while in some areas there are only a limited number of species, others, particularly certain tropical forest areas appear to have living diversity in superabundance. "Ecuador has many more plant species than the whole of Europe, which is more than 30 times as big. Madagascar has five times as many kinds of trees as the whole of temperate North America. The United States contains fewer woody species of plant than a single volcano, Mount Makiliang in the Philippines-and the entire 20 million square kilometers of the North American continent contain fewer bird species than a 2,000-square kilometre national park in Costa Rica" (Lean et. al. 133).

A species is called endemic to a region if it is found only in the specific locality and nowhere else in the world. Some areas have many endemic species. Indonesia has one sixth of the world's bird species, and nearly a quarter of them are endemic. Half of Papua New Guinea's birds, half of the Philippines' mammals, and about 80 per cent of Madagascar's plants are unique to them. Many islands have unique endemic species because of their evolutionary isolation, which are often exceedingly vulnerable because they have small populations easily wiped out by a single disaster.

"Tropical rainforests contain the greatest diversity of species; the US National Academy of Sciences reports that a typical patch, just 10 kilometers square, contains as many as 1,500 species of flowering plant, up to 750 species of tree, 400 different types of bird, 150 butterfly species, 100 kinds of reptile, and 60 species of amphibian. Insects are so abundant that no-one has yet been able to count them, but the Academy estimates that there may be as many as 42,000 in a single hectare" (Lean et. al. 133). The Amazonian rainforest helps to make South America the richest continent for wildlife and for biodiversity generally. It covers an eighth of the world's land surface, but harbours around a third of the world's birds diversity of species. Some local hotspots can contain comparable diversity to whole temperate habitats.

Coral reefs are the rainforests of the oceans; the Great Barrier Reef, for example, contains more than 3,000 animal species. "The rainforests' nearest rivals on land are areas with a Mediterranean climate-such as coastal California, the southern part of Western Australia, and the Mediterranean basin itself. These lack the rainforests' diversity of large animals, but have a huge number of endemic plant species" (Lean et. al. 133).

The Impact on Diversity of a Changing World

Climate and geography play a part in determining why some areas have so many more species than others. Areas with high temperatures and rainfall and little seasonal variation-like tropical rainforests and coral reefs-can support many more species than cold, dry places with distinctly different seasons. History also plays a role in determining why some habitats have more species. When areas became isolated from each other, as a result of continental drift, mountain formation, ocean in welling or drying out of large rainforests into smaller islands, their animal and plant life evolves in different ways.

The longer an area is isolated, the more distinct and different its inhabitants are likely to become. The best examples are islands and super-islands, such as Madagascar and Australia, with highly distinct fauna and flora, but a similar explanation has been proposed for the high diversity of the Amazon involving insular dry periods during ice ages and re-integration of these island forest sanctuaries during warmer wetter epochs.

The theory of continental drift suggests that all the major land areas were once joined in a single "supercontinent", Pangea. Between 200 and 80 million years ago this broke up, first into two land masses, Laurasia and Gondwanaland-and then eventually into the modern continents. Australia and Antarctica broke off relatively early, while Madagascar probably did not separate entirely until about 60 million years ago. New Zealand has been isolated so long it contains no land mammals. As these land masses separated they carried plants and animals with them. Some of the ancient forests of Gondwanaland are still detectable scattered as ancient beech forest stands.

As evolutionary paths begin to diverge, different species form, filling the available ecological niches. For example, Madagascar and Gondwanaland probably had very similar kinds of primitive primates when they separated. "However, over vast periods of time, Madagascar's primates (sheltered from the fierce competition that species still faced on the mainland) developed into lemurs, lower primates found nowhere else. Gondwanaland's primates, subjected to greater pressures, evolved into higher forms-including modern monkeys, apes and, ultimately, man. Madagascar has more than 6,000 unique flowering plants and half the world's species of chameleons are endemic to the island." (Lean et. al. 133)

Similarly in New Zealand many of the niches usually filled by mammals have been adopted by flightless birds. Australia's unique array of species evolved similarly in isolation from the rest of the world. Antarctica probably set out on a similar path, but became a snow-bound pristine wasteland with a few specialized species.

The Rich Endemism of Isolated Islands

Some islands have never been attached to the continents. Often volcanic, they start out as sterile outcrops of rocks, but later become colonised some even ending as low-lying atolls almost submerged and covered by their own biota. Their animal and plant life consists entirely of species which have colonized them from outside. Birds, bats and winged insects are obvious examples, as are fungi and plants with seeds able to blow in the air, resist the ravages of the sea, or be carried in the digestive tracts of birds or bats. Although only a small proportion of genetic 'landfalls' form other regions survive, over time, those that do evolve into species unique to their islands. "Almost 900 species of bird-10 per cent of the world's total-have a range of only one island (Lean et. al. 133). Old, isolated islands such as the Hawaii and the Galapagos Islands in the Pacific have relatively low natural species diversity compared with continents, although their endemic species may be quite unusual. Many of those present will be endemic. Young oceanic islands, and those like coral atolls, which are small and frequently flooded by the sea, often have very limited diversity, except for their marine life. A tropical grouper in a coral reef.

Plundered Coral Reefs face Potential Extinction by 2050. Reef depletion is a global problem. Clive Wilkinson, the coordinator of the Global Coral Reef Monitoring Network, said about 10% of the coral reefs worldwide had disappeared. Coral mining over-fishing pollutions and damage caused by boats and anchors were ruining the reefs. 30% of the world's coral reefs are in critical shape and may die within the next 10 to 20 years and an additional 30% are coming under sustained attack (New Scientist 19 Dec 97). Groupers and other natural predators are disappearing from Great Chargos Bank south of the Maldives because of illegal fishing from Sri Lanka. This is predicted to "cause the collapse of the reef ecosystem" (New Scientist 18 Oct 1997)

Islands on the Land

Any ecosystem or habitat surrounded by a different one is a biospheric island for the species which live there and similar mechanisms of evolution and immigration hold. "Flower-rich areas in Mediterranean climates are such ecological islands, since they have been separated from each other by enormous areas with quite different habitats for millions of years. They support very diverse flora, with a high percentage of endemic species. The same applies to isolated mountainous regions in the tropics, such as the highlands of Ethiopia, Cameroon and the eastern side of the rift valley in central Africa, which between them support a high proportion of the rare species of Africa" (Lean et. al. 133). Similarly, the rift valley lakes in Africa are isolated from each other and each has evolved it own highly diverse kinds of fish as noted below in the Chichlids.

ENVIRONMENTAL COLLABORATION AND DEVELOPMENT

Collaborative or cooperative approaches to environmental and natural resource management provide potential solutions to the dilemma of the environment development tradeoff. These approaches rely on positive incentives and partnership arrangements. Over the past decade the term social capital has received considerable attention from scholars in a variety of fields. Social capital is valuable because it provides resources to solve problems of coordination and cooperation, reduces transaction costs, and facilitates the flow of information between and among individuals in community or organization. Similarly, Putnam (1993) argues that social capital makes collective works easier and, ultimately, facilitates economic and community development

The concept of social capital has become increasingly popular in a wide range of social science disciplines, but there is a lack of consensus on the meaning of term. In social science research "social capital" is used in vastly different ways. Critics have characterized research examining the impacts of social capital as '"casual empiricism", because it lacks of an obvious link between theory and measurement (Durlauf, 1999). In order to better understand how social capital can help state and local governments reconcile environmental and development goals, we systematically define and classify social capital based on its scope and form. This allows us to identify different "types" of social capital that can shape collaboration and partnership among actors concerned with environment and economic development.

The Forms of Social Capital

The Uphoff (2000) suggested two dimension of social capital—structural and cognitive. Structural forms of social capital concern the roles, rules, procedures, and networks that facilitate information sharing, and collective action and decisionmaking through established roles, social networks and other social structures supplemented by rules, procedures, and precedents.

As such, it is a relatively objective and externally observable construct. Cognitive social capital refers to shared norms, values, trust, attitudes, and beliefs. It is therefore a more subjective and intangible concept (Uph off, 2000). Landry, Amara, and Lamari also classify two form of social capital: Structural and Cognitive. They measure three type of structural social capital: Network capital, Relationship capital, and Participation capital. Cognitive social capital was measured by trust capital. Krishna (2000) makes a similar distinction between institutional capital and relational capital.

The structural (Institutional) dimension of social capital includes rule of law, formal institutions and organization structures, but it also encompasses the overall pattern of relationships in an organization and its included network. This conceptualization is similar to Granovetter's (1973) notion of weak ties.

The relational dimension of social capital concerns the nature of connections between individuals. It is characterized by levels of trust, shared norms and perceived obligation, and sense of mutual identification. This conceptualization of relational social capital is similar to Granovetter's (1973) notion of strong ties. Likewise, Feiock and Tao (2002) distinguish endogenous and exogenous social capital, and examine their effects on the regional economic development partnership as one form of collective action.

The Scope of Social Capital

The scope of social capital ranges from the micro to the macro level. Analysis of social capital at the micro level is usually associated with face-to face interaction between and among individuals (Turner, 1999), and those features of horizontal relationship, such as networks of indivi duals or households, and the associated norms and trust, that generat e externalities for the community as a whole. James C oleman (1990) includes vertical as well as horizontal associations and behavior within and among organizations by expanding the unit of ob servation and introducing a vertical component to social capital.

A macro-view of social capital includes the social and political environment that shapes social structure and enables norms to develop. This view includes the most formalized institutional relationships and structures, such as the rule of law, the political regime, the court system, and civil and political liberties. This focus on institutions draws on the work of Mancur Olson (1982) and Douglas North (1990), who have argued that such institutions have a significant effect on the pattern and rate of economic development.

The phenomena related with the micro and macro level conceptualizations are complementary and their coexistence maximizes the waves of social capital on economic and social outcomes. For example, macro institutions can provide an enabling environment in which local associations can develop and flourish; local associations can sustain regional and national institutions and add a measure of stability to them.

A Typology of Social Capital

Whether at the micro or macro level, social capital exerts its influence on development as a result of the interactions between two distinct types of social capital—structural and cognitive.

Cooperation and coordination among neighbours can be based on a personal cognitive bond that may not be reflected in a formal structural arrangement.

Similarly, the existence of a community association does not necessarily testify to strong personal connections among its members, either because participation in its activities is not voluntary or because its existence has outlasted the external factor that led to its creation. Social interaction can become capital through the persistence of its effects, which can be ensured at both the cognitive and structural level.

We craft a typology of types of social capital framework based on these two key dimensions: its scope and its form. The framework treats social capital as a genuine asset that requires investment to accumulate and that generates a stream of benefits.

Ideally empirical investigation of social capital would examine and measure all four quadrants. Empirical work had generally focused on one or at most two of these quadrants. The most extensive work has been on micro level institutions or norms. Recent work (Park 2003) has used confirmatory factor analysis to empirically isolate these dimensions.

Social Capital, Environmental Collaboration

Differentiating the types of social capital may help us understand how some state and local governments are able to overcome tradeoffs between environmental and economic gains. Lubell and Scholz (2001) suggest that reciprocity in relationships among governmental and non-governmental actors and lengthy time horizons are necessary to achieve sustainable development and to overcome collective action problems in environmental management.

By extending these arguments, we contend that overcoming tradeoffs between developmental and environmental concerns requires: 1) participation in democratic political institutions; 2) social mechanisms to resolve conflicts from unharmonious development; and 3) information sharing for the diffusion of innovations. Each of these is facilitated by social capital in the community.

Specific types of social capital influence collective action and economic performance. Any form of capital-material or nonmaterial-represents an asset or a class of assets that produces a stream of benefits. The stream of benefits from social capital-or the channels through which it influences development-includes several associated elements. First, Cognitive social capital at the micro level (i.e., endogenous social capital) such as trust, shared norms, and informal sanction reduce transaction costs. Reputations built through trust and reciprocity reduce information, monitoring, and enforcement costs and thus facilitate cooperation and collective action.

Second, Structural social capital such as associations, networks, and institutions provide an informal and formal framework to organize information sharing, coordination of activities, and collective decision-making. Participation by individuals in social networks increases the availability of information and lowers its cost. This information, especially if it relates to such things as new "green" technologies can play a critical role in increasing the returns from economic production while mitigating adverse environmental consequences.

Participation in local networks and attitudes of mutual trust make it easier for a group to reach collective decisions and implement collective action. Since property rights are often imperfectly developed and applied, collective decisions on how to manage common resources are critical to maximizing their use and yield. Finally, networks and attitudes reduce opportunistic behavior by community members. In settings where a certain behavior is expected from individuals for the benefit of the group, social pressures and fear of exclusion can induce these individuals to provide the expected behavior by reducing transaction costs and encouraging innovation.

Social capital contributes to sustainable economic development and growth by reducing conflict and the transaction costs of environmental management and by facilitating information sharing and the diffusion of innovation. Environmental governance systems based on partnership provide one mechanism to exploit existing social capital in its various forms and generate additional social capital resources.

VALUES ASSOCIATED WITH BIODIVERSITY

The following categories summarise the different values people and society place on biodiversity:

Direct utilitarian value

Biodiversity is consumed by humans as food and is used to feed stock. It provides materials such as timber and fibre, medicines, chemicals and genetic material.

Indirect utilitarian values

Indirect utilitarian values include the maintenance of 'ecosystem services' or important ecological processes. Examples include maintaining water quality in catchments, moderating atmospheric processes or weather, conserving the structure or fertility of soil, maintaining coastal function, assimilating or removing wastes from water or soil, maintaining evolutionary potential in ecosystems, sequestering carbon emissions, cycling of nutrients, pest control, and pollination of crops.

Aesthetic and recreational values

Biodiversity has aesthetic and recreational uses for humans, both in the form of specific taxa such as flowers, birds, trees or whales, and as components of natural or semi-natural landscapes such as the Great Barrier Reef and the wetlands of Kakadu National Park.

Scientific and educational values

Scientific discovery can lead to the development of utilitarian values. It will often be through scientific research, other forms of investigation and learning about community or Indigenous knowledge that such uses will be recognised. Also, the variety of life is of educational value across a wide variety of subjects and disciplines (e.g. biology, biochemistry, ecology, genetics and agronomy).

Intrinsic, spiritual and ethical values

Various cultural and religious systems (e.g. Aboriginal and Torres Strait Islander people) place value on components of biodiversity. Also, there is the ethical position that non-human forms of life have intrinsic value and a right to exist independent of any use to humans.

Future or 'option' values

For all of the above values, there is the added dimension of keeping options open for the future. We are uncertain as to what species and populations are crucial to ecosystem services, or the actual significance of some of these services. Similarly, there may be uses for species or genetic diversity yet to be discovered, such as for food or medicine. And, if the values held in society change as they have in the past, then what is viewed as unimportant now may be more highly valued in the future. All the values identified above are evident in Australian society, and many individuals will value biodiversity for more than one of these reasons. Perhaps the most important change in understanding in the long term has been the recognition of the reliance of biodiversity on functioning ecosystems, and its role in maintaining ecological processes. This recasts biodiversity science, policy and management in important ways. Managing just a few species and protecting a small selection of natural areas is not sufficient to protect Australia's biodiversity.

Another major and continuing change is the attention being paid to *indirect* (or underlying) as well as *direct* (or proximate) causes of biodiversity loss. For example, land clearing by farmers is a direct cause of biodiversity loss in Australia. The indirect causes lie in the social, institutional and economic settings that influence farmer behaviour and farm profitability. This includes the information available to landholders, economic conditions affecting rural industries and perverse incentives encouraging clearance.

This shift in emphasis deepens our understanding of the processes of biodiversity loss and allows more sophisticated policy responses. In the land clearance example, strict regulation is invited by the direct cause, whereas understanding the indirect cause invites the use of incentive mechanisms, forward planning, information provision and other approaches.

ATTITUDES TOWARDS BIODIVERSITY

In the human history of Australia, changing values, scientific knowledge and cultural understanding have altered the way we perceive and interact with the natural environment generally and with biodiversity in particular.

The following six phases offer a simplified summary of this process, characterising six different sets of attitudes. While the emphasis has changed over time, these phases reflect attitudes toward the Australian environment and biodiversity that are, to some extent, current.

Indigenous

The Indigenous peoples of Australia have valued and utilised components of biodiversity for at least 60 000 years. Indigenous culture and practices have developed in a dynamic relationship with the environment. For example, the use of fire and hunting of animals helped shape the terrestrial environment.

Today, Australians live in a cultural landscape that incorporates a diversity of manifestations of the interactions between humans and the natural environment over the last 60 000 years.

While 'western science' is still dominant in terms of the way non-Indigenous people view, describe and classify our flora and fauna, there are an increasing numbers of examples where Indigenous ecological knowledge is being accepted on an equal basis (Kakadu Board of Management and Parks Australia 1998).

Exploration and contact

Australia's flora and fauna were often a focus, but mostly in terms of their possible value for colonial trade or of their scientific interest and peculiarity.

Exploitative pioneering

The environment was viewed as feedstock for colonial economic and trade development. There was little in the way of specific law or management aimed at the protection of native species. Acclimatisation Societies were set up so the early settlers could make the environment more like 'home' and brought in some plants and animals that ended up becoming pests.

Wise use for national development

New resource management arrangements were established that were informed by science and aimed at management of natural resources (especially forests, water and soils) to best answer human needs in both the present and future.

Modern environmentalism

The modern conservation movement, prominent from the 1960s onward, expressed new issues and values in environmental management debates. In this era, the intrinsic value of biodiversity was more widely recognised, and when laws and policies for biodiversity conservation became the norm rather than the exception.

Ecologically sustainable development

Since the early 1990s, the central organising concept used by governments to describe human interaction with the environment has been 'sustainable development', termed 'ecologically sustainable development' (ESD) in Australia. The main aims of ESD are to: integrate environmental, social and economic concerns over a long time; adopt a precautionary approach; and recognise the importance of biodiversity and ecosystem processes.

Current debates and practices still feature these different attitudes. One constant is the character of Australia's biota. The biota, special to Indigenous people for millennia, appeared fantastic and fascinating to early European observers. Beneath the superficial strangeness of kangaroos, black swans, platypus and endless eucalypts, a special quality has been increasingly realised.

ENVIRONMENTAL CHANGE AND BIODIVERSITY CONSERVATION

In the Convention on Biological Diversity signed by many member states at the Earth Summit held in Rio de Janeiro (Brazil) in 1992, explains biodiversity as follows: "Biological diversity" means the variability among living organisms from all sources including terrestrial, marine and other aquatic ecosystems and the ecological complexes of which they are part; this includes diversity within species, between species and of ecosystems.

Global Warming, Climate Change and Migration

GLOBAL WARMING: CAUSES AND RESULTING CLIMATE CHANGE

The most authoritative reports on the causes and consequences of climate change come from the IPCC, particularly its 1995 Second Assessment Report (SAR) and its 2001 Third Assessment Report (TAR). The latter report refined the findings of the first assessment, pointing out that climate change is likely to be worse and occur more rapidly than initially predicted. Here I summarize the IPCC's findings on global warming and the worldwide effects of climate change before pointing out some of the anticipated socio-economic impacts in East Asia.

According to the IPCC's TAR, there is now a collective picture, derived from an increasing body of observations, of a warming world and other changes in the Earth's climate system. The global average surface temperature increased during the twentieth century, with the 1990s and early 2000s the warmest on record.

Snow and ice cover have decreased, global average sea level has risen, and the heat content of the oceans has increased. Other aspects of climate have changed during the twentieth century, including changes in precipitation (e.g. increased heavy precipitation events) and cloud cover; fewer extreme low-temperature periods and more high-temperature periods; more frequent, persistent, and intense episodes of the El Nino ocean-warming event (and related adverse effects on weather in many areas); and an increase in areas experiencing drought and severe wet periods. Some climate related events, such as tornadoes or tropical storms, do not appear to have changed based on IPCC data, although the evidence is conflicting. The TAR also finds that emissions of GHGs from human activities are altering the atmosphere in ways that are expected to affect climate.

Human activities have increased atmospheric concentrations of GHGs (e.g. CO_2, methane, nitrous oxide, halocarbons) and their warming potential. According to the report, "atmospheric concentration of carbon dioxide (CO_2) has increased 31 percent since 1750. The present CO_2 concentration has not been exceeded during the past 420,000 years and likely not during the past 20 million years. The current rate of increase is unprecedented during at least the past 20,000 years".

Three-quarters of human-induced emissions of CO_2 over the last two decades has come from the burning of fossil fuels (e.g. coal, oil, and natural gas), with most of the remainder the consequence of land-use changes, particularly deforestation. Natural causes of climate change have been relatively small. Furthermore, models for predicting future climate are increasingly accurate and precise. While uncertainties remain, understanding of climate processes and predicted effects has improved.

According to the TAR, new and stronger evidence points to human activities as the sources of observed global warming over the last fifty years, further strengthening the SAR's conclusion that the "balance of evidence suggests a discernible human influence on global climate". Warming over the last 100 years is unlikely to have been natural, with studies showing that global warming, particularly during the last 35-50 years, most likely resulted from human activities.

Thus, the TAR concludes: "In light of new evidence and taking into account the remaining uncertainties, most observed warming over the last fifty years is likely to have been due to the increase in GHG concentrations. Furthermore, it is very likely that the 20th century warming has contributed significantly to the observed sea level rise ... and widespread loss of land ice".

Furthermore, the TAR determined that human activities will continue to shape the Earth's atmosphere throughout this century and into the future, and average global temperatures and sea levels are projected to rise. Emissions from burning fossil fuels will be the dominant source of atmospheric CO2 during this century. These emissions and those of other GHGs would have to be reduced to "a very small fraction of current emissions" to stabilize climate.

Global average temperature is projected by the IPCC to increase by 1.4-5.8 degrees Celsius during this century (more than anticipated in the SAR). This warming will occur at a rate faster than that observed in the twentieth century, "very likely to be without precedent during at least the last 10,000 years". During this century, warming is expected to occur in most areas, but it should be particularly pronounced at northern high latitudes during winter. Global mean sea level is expected to rise 0.09-0.88 metres in this century, with other very likely changes to include higher maximum temperatures and more hot days over most land areas, higher minimum temperatures and fewer cold days over most land areas, more intense precipitation events over many areas, increased summertime continental drying and drought over mid-latitude continental interiors, and more severe storms over some areas.

Ecological and Socio-economic Impacts of Climate Change

The ecological and socio-economic impacts of climate change are likely to be very significant and often painful. The TAR's findings on these impacts include the following: Regional changes in climate have already affected many physical and biological systems, with temperature increases being the most proximate cause.

Observed changes in regional climate have occurred in terrestrial, aquatic, and marine environments, and effects have included shrinking glaciers, thawing permafrost, reduced periods in which lakes and rivers are frozen, longer mid- and high-latitude growing seasons, shifts in animal and plant ranges to higher latitudes and altitudes, declines in populations of some animals and plants and reduced egg-laying in some birds, and insects populating new areas.

It appears that some social and economic systems have already been affected by increased floods and drought, but separating these ecological events from socio-economic factors is difficult.

The TAR shows that many human systems are sensitive to climate change, including water resources, agriculture, coastal zones and marine fisheries, settlements, energy, industry, financial services (e.g. insurance industries affected by increased claims), and human health.

Adverse impacts of climate change include reduced crop yields in most tropical and sub-tropical regions; decreased water availability in many water-scarce areas, especially the sub-tropics; more people exposed to increased mortality from heat stress and vector-borne diseases like cholera; widespread increase in flood risk from rising sea levels; and increasing demand for energy to cool areas affected by higher summer temperatures.

Some impacts may be positive, such as increased crop yields in some mid-latitude areas; potentially more timber if forests are managed appropriately (although increased pests could more than offset this); increased water availability for some water scarce areas; lower winter mortality in traditionally cold areas; and reduced winter demand for energy due to higher winter temperatures.

Many of the risks are unclear, and there is substantial potential for "large-scale and possibly irreversible impacts" from changing ocean currents, melting ice sheets, accelerated global warming due to atmospheric feedback effects, and so forth. In addition to efforts to mitigate climate change, the TAR argues that adaptation is a necessary strategy.

However, those people and societies with the least resources are most vulnerable because they are least able to adapt. Projected warming may result in a mixture of economic gains and losses for developed countries, but developing countries can expect mostly losses: "The projected distribution of economic impacts is such that it would increase the disparity in well-being between developed countries and developing countries," with the disparities increasing the greater the temperature increases. The upshot is that "More people are projected to be harmed than benefited by climate change," even if temperature increases are limited.

The TAR is not restricted to scientific and economic assessments. It argues that international justice and equity are important considerations when addressing climate change: "Inclusion of climatic risks in the design and implementation of national and international development initiatives can promote equity and development that is more sustainable and that reduces vulnerability to climate change".

Global Warming and Climate Change Impacts in East Asia

Clearly, the global effects of climate change are potentially major, and will likely lead to many adverse consequences, difficult choices, and expensive adaptation measures for much of the world's population. The countries of East Asia will not be immune to these changes, and in most cases will be among the worst affected due to their vulnerable geographies and economies.

Effects may not always be adverse, but even if they are not they will likely increase unpredictability and require adaptation. What are the expected impacts of climate change in East Asia? Several research reports have anticipated the effects of climate change for the region. Some of their findings are summarized here to convey the scale and nature of the potential changes.

According to a 1997 report from the IPCC on anticipated regional impacts of climate change, temperate Asia (including Japan, the Koreas, and most of China) has experienced an average annual temperature increase of more than 1 degree Celsius in the last century, mostly since the 1970s, with substantial warming expected in this century. Rainfall is expected to change in the area, with substantial declines expected in most of China (notably northern provinces). Permafrost in northeast China is expected to disappear (with release of methane, thus adding GHGs to the atmosphere) and glaciers will melt. Northern China is particularly vulnerable to expected changes in rainfall, exacerbating existing water shortages.

The area is likely to experience changing agricultural yields, with many crops likely to see reductions and a northward movement of crop zones and anticipated shortages of roundwood (partly due to increased demand). Delta coastlines in China "face severe problems" from sea-level rise, which will include salt water intrusion into aquifers. Japan will not be immune; already many parts of major coastal urban areas, with millions of residents, are below the mean high-water mark.

Providing protection for only some of these cities will cost tens of billions of dollars. Japan's beaches, which comprise about a quarter of its coastline, will be subject to erosion - and over half of existing beaches may disappear. Additionally, heat-related deaths throughout temperate Asia may increase sevenfold by the middle of this century.

The potential effects of climate change for tropical Asia (encompassing Southeast Asia) are also described in the IPCC's 1997 regional report. It points out that the region already suffers from increasing pollution, land degradation, and all manner of environmental problems resulting from rapid urbanization, industrialization, and economic development. Climate change will exacerbate these problems. In this area, mean temperatures have already gone up by 0.3-0.8 degrees Celsius over the last 100 years.

Forest cover will change as a consequence, possibly increasing, and forest types may change. Changes in evaporation and rainfall are likely to have detrimental effects on freshwater wetlands. Coastal areas will be most greatly affected by sea-level rise and increased ocean temperatures (the latter possibly preventing coral reefs from keeping up with sea-level rise). Mangrove and tidal wetlands will have difficulty adapting due to bordering infrastructure and human activities. Greater erosion, coastal flooding, and salinization of fresh water sources are probable. Delta regions of Southeast Asian countries are particularly vulnerable, and throughout this area several million people could be displaced by sea-level rise.

The costs of responding to the impacts of rising seas, in the words of the IPCC, could be immense. Glaciers feeding the area's rivers will melt, and there may be yearly reductions - albeit between seasonal flooding - in the flow of snow-fed rivers, adversely affecting agriculture, hydropower generation, and urban water supplies.

Agriculture will probably suffer (despite CO_2 fertilization) from temperature and moisture changes and possibly from increased pests, affecting, for example, wheat, rice, and sorghum crops (although much uncertainty, confounding planning, will obtain). According to the IPCC report, poor rural populations depending on traditional forms of agriculture or living on marginal lands are especially vulnerable. Increased vector-borne diseases such as dengue, malaria, and schistosomiasis will adversely affect human health in this area.

A 1999 report on climate change impacts prepared by Britain's Climatic Research Unit summarized many potential impacts for some of the countries of East Asia. In China, temperature increases are predicted to be greatest over northern areas, with changes in precipitation and threats to biodiversity. In Indonesia, forest fires are predicted to increase and endangered species may be threatened.

In Japan, heat waves will increase in frequency and intensity, coastlines and coastal infrastructure will be harmed, and reefs will be stressed. In the Philippines, rainfall will increase during the wet season and decrease during the dry season, reefs will suffer from warming water, and potentially millions of people will be threatened by sea-level rise.

Von Hippel has summarized a few of the possible impacts of climate change in Northeast Asia: pressure on agricultural resources and accelerated desertification leading to cross-border migrations, particularly from China to Russia; adverse climatic effects on North Korea's food production, possibly increasing military pressure on South Korea and creating economic burdens for reunification; increased demand for air conditioning, leading to higher fuel consumption and hence more local and regional air pollution (and adding still further to GHG emissions); salinization of breeding grounds for fish from sea-level rise, leading to reduced fishery yields that could exacerbate conflicts over marine resources; additional oil pollution (from shipments of oil imports) that may strain relations among countries sharing marine resources and shipping lanes; and increased economic costs from natural disasters like catastrophic storms, straining emergency and disaster relief resources in the region.

The 2001 TAR assessment of vulnerability in Asia shows that the region is potentially more susceptible to climate change than are some other regions of the world. It concludes that the developing countries of Asia are highly vulnerable to climate change, and their adaptability is low. (Developed countries of the region (e.g. Japan) are of course less vulnerable because they are more able to adapt to climate change.)

Floods, forest fires, cyclones, droughts and other extreme events have increased in temperate and tropical Asia. The TAR anticipates that while agricultural productivity could increase in northern parts of Asia, food security would suffer in arid, tropical, and temperate Asia due to reduced agricultural and aquaculture productivity from warmer water, sea-level rise, floods, droughts, and cyclones.

Water availability may decrease in arid and semi-arid Asia and possibly increase in northern Asia, and increased incidence of vector-borne diseases and heat-stress will threaten human health. Temperate and tropical Asia should anticipate increased rainfall and floods, and sea-level rise and more intense storms could "displace tens of millions of people in low-lying coastal areas of temperate and tropical Asia". Some parts of Asia will see climate change effects on transport, increased demand for energy, and adverse impacts on tourism. Land-use and land-cover changes will threaten biodiversity, and sea-level rise will adversely impact coral reefs and mangrove areas that are important for fisheries.

What comes from these (and other) reports on the impacts of climate change in East Asia is that many of the effects will be felt most by - and be most painful for - the developing countries of the region. They are generally more vulnerable and least able to cope due to poverty and existing environmental problems and resource scarcities.

A very large number of people throughout East Asia live in low-lying coastal regions, and they are threatened by sea-level rise, land subsidence, inundation of fresh water aquifers by salt water, and more frequent and violent storms from climate change. Island countries such as Indonesia and the Philippines are especially vulnerable to climate change effects.

They can expect freshwater shortages and damage to coastal areas and adjacent infrastructure, with concomitant adverse effects on tourism. (Indeed, in extreme cases it may one day be necessary for some small-island states to abandon their territory altogether. Representatives from these countries have for some time argued that they are *already* feeling the effects of rising oceans.

Other poorer countries in the region are vulnerable. For example, the World Bank reported that Chinese research has estimated that a 1-metre rise in sea level would inundate 92,000 square kilometres of China's coast, displacing 67 million people (and more as population increases). According to one assessment, future climate change will reduce soil moisture in China, particularly in the north, and this will increase the demand for agricultural irrigation, which will in turn add to existing severe water shortages.

In short, "Possible impacts of climate change on Chinese agriculture could be highly disruptive ...". Already vulnerable, China may also see greater weather extremes, including droughts in the north and floods in the south, and heat stroke and death will increase, as may occurrences of malaria, dengue fever, and other diseases.

An ever-growing body of research shows that climate change will (and probably has already) adversely affected human health, and this is particularly true of East Asia. Southeast Asia is especially vulnerable to anticipated increasing incidence of vector-born diseases. Hotter weather will increase heat-related mortality in the region, as indicated by historical studies from China showing a strong correlation between peak summer temperatures and death rates.

But even the developed countries and regions of East Asia are unlikely to avoid harm from climate change. For example, while Japan's coastlines are not as vulnerable as those of China, the Philippines, and other countries, it is reasonable to expect that it will suffer costly damage from sea-level rise, associated storm surges, and adverse weather, and it has direct interests in the health of surrounding seas and indirect interest in what happens throughout the region.

By way of example, recently reduced fish catches by Japanese fisherman have been attributed to changes in underwater currents triggered by global warming. And there will be adverse impacts for Japan's biodiversity, forests, agriculture, wetlands, and water systems, as well as for infrastructure and human health. According to the government, climate change effects have already become visible in Japan. For these and other reasons, Japan supports the climate change regime and the Kyoto Protocol- despite its greater ability to cope with climate change compared to its neighbours.

GLOBAL WARMING AND CLIMATE CHANGE

Unchecked global warming could affect most terrestrial ecoregions. Increasing global temperature means that ecosystems will change; some species are being forced out of their habitats because of changing conditions, while others are flourishing. Secondary effects of global warming, such as lessened snow cover, rising sea levels, and weather changes, may influence not only human activities but also the ecosystem.

For the IPCC Fourth Assessment Report, experts assessed the literature on the impacts of climate change on ecosystems. Rosenzweig et al. concluded that over the last three decades, human-induced warming had likely had a discernable influence on many physical and biological systems. Schneider et al. concluded, with very high confidence, that regional temperature trends had already affected species and ecosystems around the world. With high confidence, they concluded that climate change would result in the extinction of many species and a reduction in the diversity of ecosystems.

- *Terrestrial ecosystems and biodiversity*: With a warming of 3°C, relative to 1990 levels, it is likely that global terrestrial vegetation would become a net source of carbon. With high confidence, Schneider et al. concluded that a global mean temperature increase of around 4°C by 2100 would lead to major extinctions around the globe.

- *Marine ecosystems and biodiversity*: With very high confidence, Schneider et al. concluded that a warming of 2°C above 1990 levels would result in mass mortality of coral reefs globally.
- *Freshwater ecosystems*: Above about a 4°C increase in global mean temperature by 2100, Schneider et al. concluded, with high confidence, that many freshwater species would become extinct.

Studying the association between Earth climate and extinctions over the past 520 million years, scientists from the University of York write, "The global temperatures predicted for the coming centuries may trigger a new 'mass extinction event', where over 50 per cent of animal and plant species would be wiped out." Many of the species at risk are Arctic and Antarctic fauna such as polar bears and Emperor Penguins. In the Arctic, the waters of Hudson Bay are ice-free for three weeks longer than they were thirty years ago, affecting polar bears, which prefer to hunt on sea ice.

Species that rely on cold weather conditions such as gyrfalcons, and Snowy Owls that prey on lemmings that use the cold winter to their advantage may be hit hard. Marine invertebrates enjoy peak growth at the temperatures they have adapted to, regardless of how cold these may be, and cold-blooded animals found at greater latitudes and altitudes generally grow faster to compensate for the short growing season.

Warmer-than-ideal conditions result in higher metabolism and consequent reductions in body size despite increased foraging, which in turn elevates the risk of predation. Indeed, even a slight increase in temperature during development impairs growth efficiency and survival rate in rainbow trout. Rising temperatures are beginning to have a noticeable impact on birds, and butterflies have shifted their ranges northward by 200 km in Europe and North America. Plants lag behind, and larger animals' migration is slowed down by cities and roads. In Britain, spring butterflies are appearing an average of 6 days earlier than two decades ago.

A 2002 article in Nature surveyed the scientific literature to find recent changes in range or seasonal behaviour by plant and animal species. Of species showing recent change, 4 out of 5 shifted their ranges towards the poles or higher altitudes, creating "refugee species". Frogs were breeding, flowers blossoming and birds migrating an average 2.3 days earlier each decade; butterflies, birds and plants moving towards the poles by 6.1 km per decade. A 2005 study concludes human activity is the cause of the temperature rise and resultant changing species behaviour, and links these effects with the predictions of climate models to provide validation for them. Scientists have observed that Antarctic hair grass is colonizing areas of Antarctica where previously their survival range was limited.

Mechanistic studies have documented extinctions due to recent climate change: McLaughlin et al. documented two populations of Bay checkerspot butterfly being threatened by precipitation change. Parmesan states, "Few studies have been conducted at a scale that encompasses an entire species" and McLaughlin et al. agreed "few mechanistic studies have linked extinctions to recent climate change." Daniel Botkin and other authors in one study believe that projected rates of extinction are overestimated.

Many species of freshwater and saltwater plants and animals are dependent on glacier-fed waters to ensure a cold water habitat that they have adapted to. Some species of freshwater fish need cold water to survive and to reproduce, and this is especially true with Salmon and Cutthroat trout. Reduced glacier run-off can lead to insufficient stream flow to allow these species to thrive. Ocean krill, a cornerstone species, prefer cold water and are the primary food source for aquatic mammals such as the Blue Whale. Alterations to the ocean currents, due to increased freshwater inputs from glacier melt, and the potential alterations to thermohaline circulation of the worlds oceans, may affect existing fisheries upon which humans depend as well.

The white lemuroid possum, only found in the mountain forests of northern Queensland, has been named as the first mammal species to be driven extinct by global warming. The White Possum has not been seen in over three years. These possums cannot survive extended temperatures over 30 °C which occurred in 2005. A final expedition to uncover any surviving White Possums is scheduled for 2009.

Forests

Pine forests in British Columbia have been devastated by a pine beetle infestation, which has expanded unhindered since 1998 at least in part due to the lack of severe winters since that time; a few days of extreme cold kill most mountain pine beetles and have kept outbreaks in the past naturally contained. The infestation, which has killed about half of the province's lodgepole pines is an order of magnitude larger than any previously recorded outbreak and passed via unusually strong winds in 2007 over the continental divide to Alberta.

An epidemic also started, be it at a lower rate, in 1999 in Colourado, Wyoming, and Montana. The United States forest service predicts that between 2011 and 2013 virtually all 5 million acres of Colourado's lodgepole pine trees over five inches in diametre will be lost.

As the northern forests are a carbon sink, while dead forests are a major carbon source, the loss of such large areas of forest has a positive feedback on global warming. In the worst years, the carbon emission due to beetle infestation of forests in British Columbia alone approaches that of an average year of forest fires in all of Canada or five years worth of emissions from that country's transportation sources.

Besides the immediate ecological and economic impact, the huge dead forests provide a fire risk. Even many healthy forests appear to face an increased risk of forest fires because of warming climates. The 10-year average of boreal forest burned in North America, after several decades of around 10,000 km^2 has increased steadily since 1970 to more than 28,000 km^2 annually.

Though this change may be due in part to changes in forest management practices, in the western U.S., since 1986, longer, warmer summers have resulted in a fourfold increase of major wildfires and a sixfold increase in the area of forest burned, compared to the period from 1970 to 1986.

A similar increase in wildfire activity has been reported in Canada from 1920 to 1999.

Forest fires in Indonesia have dramatically increased since 1997 as well. These fires are often actively started to clear forest for agriculture. They can set fire to the large peat bogs in the region and the COreleased by these peat bog fires has been estimated, in an average year, to be 15% of the quantity of COproduced by fossil fuel combustion.

Mountains

Mountains cover approximately 25 per cent of earth's surface and provide a home to more than one-tenth of global human population. Changes in global climate pose a number of potential risks to mountain habitats. Researchers expect that over time, climate change will affect mountain and lowland ecosystems, the frequency and intensity of forest fires, the diversity of wildlife, and the distribution of water.

Studies suggest that a warmer climate in the United States would cause lower-elevation habitats to expand into the higher alpine zone.

Such a shift would encroach on the rare alpine meadows and other high-altitude habitats. High-elevation plants and animals have limited space available for new habitat as they move higher on the mountains in order to adapt to long-term changes in regional climate.

Changes in climate will also affect the depth of the mountains snowpacks and glaciers. Any changes in their seasonal melting can have powerful impacts on areas that rely on freshwater run-off from mountains. Rising temperature may cause snow to melt earlier and faster in the spring and shift the timing and distribution of run-off. These changes could affect the availability of freshwater for natural systems and human uses.

THE ETHICAL DIMENSIONS OF CLIMATE CHANGE

The major outcome of this meeting was the Buenos Aires Declaration on the Ethical Dimensions of Climate Change.

Objectives

The programme on the Ethical Dimensions of Climate Change seeks to:

- Facilitate express examination of ethical dimensions of climate change particularly for those issues entailed by specific positions taken by governments, businesses, NGOs, organizations, or individuals on climate change policy matters;
- Create better understanding about the ethical dimensions of climate change among policy makers and the general public;
- Assure that people around the world, including those most vulnerable to climate change, participate in any ethical inquiry about responses to climate change;
- Develop an interdisciplinary approach to inquiry about the ethical dimensions of climate change and support publications that examine the ethical dimensions of climate change;
- Make the results of scholarship on the ethical dimensions of climate change available to and accessible to policy makers, scientists, and citizen groups;
- Integrate ethical analysis into the work of other institutions engaged in climate change policy including the Intergovernmental Programme on Climate Change and the Conference of the Parties to the United Nations Conference on Climate Change.

Position

Given the severity of impact to be expected and given the likelihood that some level of important disruptions in living conditions will occur for great numbers of people due to climate change events, this group contends that there is sufficient convergence among ethical principles to make a number of concrete recommendations on how governments should act, or identify ethical problems with positions taken by certain governments, organizations, or individuals. Facts about climate change and fundamental human rights provide the starting point for climate ethics.

World Climate Report

World Climate Report, a newsletter edited by Patrick Michaels, was produced by the Greening Earth Society, a non-profit organization created by the Western Fuels Association. Early editions were document based; it was then transferred to a web-only format, having ceased publication as a physically based report with Volume 8 in 2002.

World Climate Report presents a scientific skeptical view of populist anthropogenic-driven mass global climate change, or as it describes, 'Global Warming Alarmism'. However, it does not reject the concepts of global climate change or greenhouse theory in general attempting to engender itself as giving a well balanced and scientific view of the sources.

WCR says of itself:

- World Climate Report, a concise, hard-hitting and scientifically correct response to the global change reports which gain attention in the literature and popular press. As the nation's leading publication in this realm, World Climate Report is exhaustively researched, impeccably referenced, and always timely. This popular biweekly newsletter points out the weaknesses and outright fallacies in the science that is being touted as "proof" of disastrous warming. It's the perfect antidote against those who argue for proposed changes to the Rio Climate Treaty, such as the Kyoto Protocol, which are aimed at limiting carbon emissions from the United States... World Climate Report has become the definitive and unimpeachable source for what nature now calls the "mainstream skeptic" point of view..

PROTECT FORESTS FROM THE PERILS OF CLIMATE CHANGE

Jerry Franklin points out that "forest management can either exacerbate or reduce the effects of climatic change on the productivity and biological diversity of northwest forestscapes." To increase the chances that we will continue to enjoy the diverse benefits we receive from northwest forests, we must maintain and enhance their ability to respond to change. The key components of such a strategy are:

- Maintain biodiversity in all its dimensions. This will be critical, because genetic diversity is like a library of possibilities that have worked well during past climate variability, representing the sum of "tools" available for the future.
- Protect intact native ecosystems where species relations have stood the test of time and remain robust;

- Provide refugia and allow species to migrate. Buffer and expand protected areas to provide connectivity along climatic gradients. Manage the entire landscape to be amenable to dispersal of native species.
- Protect streams. Cold water fish are particularly vulnerable to climate change because of increased winter flooding, reduced summer stream flow, and increased stream temperature. To mitigate expected effects on fish we should provide generous riparian buffers to help shade streams and maintain lower stream temperatures. To render streams more resilient to hydrologic extremes, such as flooding, we should manage whole watersheds to improve their ability to absorb, store, and slowly release water. This can be accomplished in part by reducing disturbance of vegetation and soils, reducing road densities, and retaining abundant woody debris.

Logging Releases Significant Amounts of Carbon

Not surprisingly, logging accelerates the transfer of carbon to the atmosphere by killing trees that would otherwise continue to capture and store carbon through photosynthesis and growth. Killing trees also stops them from pumping carbon into the soil where much of the carbon in forests is stored. Logging actually accelerates the rate of decomposition of wood via several mechanisms. By removing the forest canopy and exposing the soil to more sunlight, logging raises soil temperature which increases the rate of decay. Logging also breaks up woody material in the forest thereby decreasing the average piece size and increasing the surface area exposed to microbial decomposition. Finally, logging debris is often burned on site or as part of an industrial process.

Traditional logging also increases the risk of disturbances. Logging increases wind damage by creating exposed edges and increasing wind speeds within forest stands. Logging often increases the wildfire hazard by making the stand hotter, dryer, and windier; by moving the most flammable small fuels from the forest canopy to the forest floor (*i.e.*, logging slash) where they are more available for combustion; and by initiating the establishment of dense stands of young trees with interlocking branches (resinous fuels) close to the ground. Logging roads also increase the risk of human-caused fire ignitions and spread tree diseases like Port Orford cedar root disease that kill trees and release carbon.

Scientists estimate that a large fraction of all the carbon transferred to the atmosphere by humans has been released due to forest exploitation. In recent decades CO_2 emissions resulting from human-induced changes to forests exceed CO_2 emissions from all motor vehicle sources combined, but forest releases are less than total emissions from all uses of fossil fuels. After logging an old-growth forest, the site remains a net source of carbon for more than 20 years, and depending on the conditions, the site does not rebuild pre-logging carbon stores for a century or more. As a result of widespread clearcutting and aggressive slash burning, the Pacific Northwest has contributed huge quantities of carbon to the atmosphere.

Increase Carbon Storage in Forests

Here in the Pacific Northwest we live in the midst of a globally significant carbon pool that should be nurtured and conserved to help keep carbon out of the atmosphere. Temperate old-growth forests of the Pacific Northwest contain some of the highest amounts of biomass per acre measured anywhere in the world. About half of the dry weight of forest biomass is comprised of carbon. The latest IPCC Mitigation Report notes that "Forest-related mitigation activities can considerably reduce emissions from sources and increase CO_2 removals by sinks at low costs..." The IPCC also states that more than 1/3 of the potential mitigation available from forests is located outside the tropics and half of the forest mitigation will come from changes in forest practices, rather than simply preventing deforestation. The objectives of forest management with respect to mitigating climate change should be two-fold effort to *protect* and *restore* forests –

- Minimize the release of additional forest carbon into the atmosphere. The best way to *retain* carbon in existing forests is to protect mature and old-growth forests and roadless areas.
- Rebuild depleted carbon stores within forested landscapes. Probably the best way to *rebuild* forest carbon stores in forests is to allow forests that were previously logged or burned to regrow and become mature and old-growth forests.

There are significant complementary benefits of managing forests for carbon storage to ameliorate global climate change. If done carefully, forests managed to provide public services such as clean water, habitat for fish and wildlife, soil conservation, and an enhanced amenity-based economy will also store large amounts of carbon over time.

Forests exhibit a quality known as "ecological inertia" which recognizes that established forests are generally long-lived, resilient to disturbance, and help create conditions suitable for their own survival. This means that our northwest forests may be able to persist through some climate changes and continue to store carbon and provide other benefits, as long as they are not clearcut or severely disturbed. This implies that if we want continued carbon storage in forests that are at the edges of their suitable range we should avoid stand-replacing logging methods (such as clearcutting) and, where ecologically appropriate, we may need to strategically reduce fuels to reduce the risk of stand-replacing fire.

Such fuel reduction must be done carefully however, because excessive removal of vegetation not only compromises carbon storage in both plants and soil, but can also increase fuel loads and fire hazard. Recent fire/fuel models indicate that forest fire hazard can be managed reasonably well by treating about 20-30 per cent of the landscape in strategic locations.

Treating fuel on every acre is neither needed or desired. Logging need not be the primary tool for accomplishing fuel reduction, because non-commercial techniques, such as low-intensity prescribed fire, are available and effective.

Forest Management Recommendations

Private forestlands: Short-rotation clearcutting typically practiced by private industrial forest land-owners is probably the worst possible way to manage forests for carbon storage, because the young forests never develop large carbon stores; significant soil carbon is lost during and after clearcutting, slash disposal, and site preparation; and the resulting wood products produced have limited longevity. Where logging is expected to continue, scientists recommend that carbon release can be mitigated if forest managers:

- Allow trees to grow much longer before harvest (*i.e.*, longer rotations),
- Retain more live trees on every acre during harvest (*i.e.*, thin instead of clearcut),
- Retain more dead wood after harvest (*e.g.* protect snags, practice less intensive slash disposal and site preparation), and
- Take steps to reduce road systems and prevent soil erosion, which would help store more carbon in forest soils.

Public lands: Federal forests can help mitigate climate change if they are restored to their natural-sustainable level of biomass and biodiversity. Large stores of carbon exist within roadless areas and mature and old-growth forests on federal lands. These should be protected from harvest, while previously logged younger forests should be carefully restored to a mature and old-growth condition that has optimal biomass storage. This management approach luckily complements other highly sought-after forest values that are currently under-represented in our forests. Careful management of forests for carbon storage can help resolve ongoing controversies over forestry's impact on water quality, old-growth, roadless areas, fish and wildlife habitat, and scenic values.

Market Solutions: Given humanity's slow response to the growing evidence of human-induced climate change and its consequences, aggressive approaches such as market intervention are now needed. The debate continues on whether a carbon tax or cap-and-trade system is better, but either is better than nothing. A carbon tax system establishes the price of carbon and the market determines how much is sequestered and not emitted. In a cap-and-trade carbon market, government would determine how much total carbon can be emitted from all sources and the market would determine who is allowed to emit the carbon and at what price.

Under current international climate protocols it is possible that forest owners of the Pacific Northwest might seek compensation for storing "extra" carbon. This would reward forest managers for storing carbon that would otherwise be transferred to the atmosphere and help off-set some of the economic costs of managing forests for carbon storage. However, there are unresolved issues about how to account for the full carbon consequences of proposed forest management activities.

For instance, the Kyoto Protocol has some "perverse incentives" that could reward carbon-poor young forests at the expense of carbon-rich old forests, though this is not scientifically supported.

In contrast to the sink management proposed in the Kyoto protocol, which favours young forest stands, we argue that preservation of natural old-growth forests may have a larger effect on the carbon cycle than promotion of regrowth.... [I]ncreasing life-span of the stand, proportionally more carbon can be transferred into a permanent pool of soil carbon (passive soil organic matter or black carbon)... [R]eplacing unmanaged old-growth forest by young Kyoto stands... will lead to massive carbon losses to the atmosphere mainly by replacing a large pool with a minute pool of regrowth and by reducing the flux into a permanent pool of soil organic matter.

Carbon stored in wood products generally do not last as long as they would if left safe inside a mature tree, but we can improve the carbon storage equation by using less wood and by increasing the lifespan of wood products. It's not just American's big cars and SUVs that are a problem; it's also their increasingly large houses. We should consider policies to help reverse the national trend towards larger houses, and we should build houses that last for centuries instead of just decades.

What about Forest Fires?

We cannot avoid the fundamentally dynamic nature of forests. Fire is an unavoidable part of life in western forests and we must stop fighting a losing battle against the inevitable. Most western forests are in some ways *dependent* upon disturbances such as fire, and past fire suppression has exacerbated rather than solved the problem of fire. Our goal should not be to prevent all damage from fires, insects, etc. Fire should be allowed to operate within natural bounds, as long as it doesn't threaten public safety. Communities and property owners in forest settings must take responsibility for becoming fire resilient or fire permeable. We should maintain healthy forest habitat by allowing natural disturbance processes to operate and expect forest carbon stores to ebb and flow, while also allowing forests to grow for long periods (and store lots of carbon) in between these natural disturbances. We must take a long-term and landscape view, so that we *optimize* carbon storage at any given point in space and time in order to *maximize* carbon storage over large landscapes and long time frames.

Fuels could be reduced in forests that are significantly outside the natural range of variability, but this must be done in a strategic and limited way that protects all large fire resilient trees and spatially disconnects large expanses of excessive fuels, while retaining as much biomass as sustainably possible. Current enthusiasm for fuel reduction must be tempered with a realization that removing too much fuel makes forests hotter, dryer, and windier which increases fire hazard and increases decomposition rates, both of which counter carbon storage and other objectives. After fire, the goal should be to retain carbon on site and allow the recovering forest to grow into a mature and old-growth condition. Aggressive replanting as recommended by the timber industry is unsupported because it establishes a dense fuel-laden condition that is susceptible to drought and is soon ripe for another fire. Natural regeneration of forests leads to more diverse and less dense forests, which is preferable from a climate change perspective because the resulting habitat diversity and spatial discontinuity are more resilient to future hazards.

DETERMINES GLOBAL TEMPERATURE AND CLIMATE

Global temperature and climate are largely determined by the balance of incoming energy from the sun, minus outgoing radiation. Incoming light radiation from the sun has short-wavelengths and can readily pass through the atmosphere, but after being absorbed and re-radiated from Earth's surfaces the out-going infra-red radiation has longer wave-lengths and is less able to pass through the atmosphere. The so-called "greenhouse gases" absorb and re-radiate a portion of the outgoing long-wave radiation back towards earth, acting like a heat-trapping blanket. Even slight changes in the ratio of incoming and outgoing solar energy have significant influence on our global climate system. Even though greenhouse gasses make up less than 1 per cent of Earth's atmosphere, our global climate is quite sensitive to changes in their concentration.

Ice-core data from Greenland and Antarctica tells us that atmospheric levels of carbon dioxide (CO_2) vary somewhat predictably with cycles of ice ages and warm inter-glacial periods. The ice cores also show that atmospheric CO_2 is increasing almost 100 times faster today than during past climate cycles, and that current concentrations of CO_2 are higher than at any time in at least the last 800,000 years. Given the difficulty of rapidly changing our resource-intensive lifestyles, we'll be lucky if global atmospheric CO_2 concentration merely doubles. More likely it will go much higher before we control our appetite for fossil fuels and land exploitation.

While CO_2 is of primary concern among greenhouse gasses, there are others such as methane (CH_4) that contribute to global warming. CO_2 is unique in that is has a very long, approximately 100 year, "residence time" in the atmosphere. Concentrations of CO_2 in the atmosphere will likely remain far above "normal" for centuries, because the millions of tons of CO_2 released to the atmosphere during the agricultural revolution, the industrial revolution, and the automobile revolution will not reach a new equilibrium until biological and geophysical processes (in the oceans and on land) have a chance to capture and store most of the "extra" carbon. We have a moral obligation to leave future generations with choices and opportunities for survival. We must avoid irreversible harm to the planet's life support systems including a livable climate and function ecosystems that sustain life.

Carbon move in and out of the atmosphere

There is a fixed amount of carbon on planet earth which is distributed among several carbon reservoirs or pools in the atmosphere, biosphere, hydrosphere, and lithosphere. In the grand scheme, carbon is neither created nor destroyed but continually moves between these various pools owing to the operation of natural and human-induced processes. The root cause of global climate change is that human activity has shifted massive quantities of carbon to the atmosphere from forests, soil, and fossil deposits.

In the atmosphere carbon is stored as CO_2, methane (CH_4), and other organic compounds. Carbon moves *into* the atmosphere from decomposition of organic matter, respiration by living organisms, combustion, volcanic activity, burning fossil fuels, degassing of waterbodies, etc. Carbon moves *out of* the atmosphere via photosynthesis, rock weathering, dissolution in water, etc. All plants, including forests and many micro-organisms, use photosynthesis which takes CO_2 out of the air to build sugars that can be used by the cell to build cellulose or other complex carbon molecules that comprise plant biomass.

This process is called "primary production" and it feeds the bottom of the global food chain. Virtually all life on earth, including humans, relies directly or indirectly on photosynthesis. Most terrestrial plants share a significant portion of their photosynthate with soil organisms, a cooperative relationship that builds a large and complex underground ecosystem that also stores carbon. Plants shed dead leaves and wood which also builds carbon stores in the soil.

In the hydrosphere (e.g. the oceans) carbon is stored mostly as dissolved $CO_2$and other dissolved organic compounds that originated in some photosynthetic life form. Carbon moves *into* the ocean from the atmosphere and biosphere via dissolving of gaseous CO_2 in cold seas, leaching from soil, and input of organic matter from river systems and the biosphere. Carbon moves *out of* the ocean primarily via photosynthesis (*e.g.* phytoplankton and cyanobacteria), degassing of warm seas, and deposition in marine sediments.

In the biosphere carbon is stored as live or recently dead plants, animals, and micro-organisms both in the ocean and on land (*e.g.*, forests and soils). Forests dominate the terrestrial carbon cycle, harbouring 86 per cent of the planet's above-ground carbon and 73 per cent of the planet's soil carbon. Carbon enters *into* the biomass pool via photosynthesis, then becomes entrained and cycled through the entire global food chain. Carbon moves *out of* the biomass pool through decomposition and respiration or through deposition in long-term storage in soil or geologic and fossil deposits.

In fossil deposits, the carbon from long-dead plants and animals are stored as coal, oil, "natural gas," or kerogen. These can be thought of as both "ancient sunlight" and "ancient atmosphere." Carbon moves *into* the fossil pool via deposition and storage in low-oxygen conditions. Carbon moves *out of* fossil pool mostly via industrial exploitation and combustion.

In the non-fossil lithosphere carbon is stored in carbonate rocks such as limestone and chalk. Carbon moves *into* these geologic structures mostly through ocean deposition. A portion of the oceanic carbon is taken up to make the shells of marine organisms that fall to the deep ocean floor where they may be subducted beneath the earth's crust and end up in long-term geologic storage, *e.g.* the Cliffs of Dover. Carbon moves *out of* the lithosphere mostly via volcanic activity and human industry such as the manufacture of cement which heats limestone and releases significant quantities of CO_2.

The advent and diversity of life on earth has had a profound impact on the global carbon cycle and now plays a fundamental role in determining whether or not we have a livable climate. The abiotic carbon cycle that existed before the proliferation of life was less stable than the carbon cycle that developed after marine organisms started to make calcium carbonate shells and deposit carbon in deep storage which has helped buffer CO_2 extremes over long time scales. Scientists have found a correlation between biodiversity and levels of atmospheric CO_2 over the last 370 million years. Human activity, mostly in just the recent era, has dramatically reallocated global carbon stores from the other carbon reservoirs into the atmosphere where it can influence our climate. For example, burning fossil fuels and heating limestone to make cement move carbon from long-term fossil and geologic storage into the atmosphere. Logging kills trees - stops carbon-uptake via photosynthesis, and moves carbon from living forests and soil into the atmosphere. Land uses such as agriculture, livestock grazing, and draining swamps move carbon from the soil to the atmosphere.

Climate Change Affect the Pacific Northwest

While predicting the local *weather* is an uncertain science, *climate* prediction is actually more accurate because the focus is on large-scale trends rather than local details. We know that the planet as a whole is almost certain to become warmer on average, and scientists expect an acceleration of the hydrologic cycle as warmer temperatures lead to increased evaporation from the oceans and more transpiration from plants. However, the effects of climate change will not be uniform around the globe. Significant uncertainty remains about how global trends will express themselves regionally. Future climate in the Pacific Northwest is even more uncertain because of complex topography and uncertain changes in precipitation, but our close proximity to the moderating influence of the Pacific Ocean likely offers a slight buffer from climate extremes.

The Pacific Northwest should expect continued climate variability. Existing cycles of cool-wet winters and warm-dry summers will likely continue, though they will be superimposed on a warmer average climate. Both floods and droughts have been part of our past and will almost certainly be part of our future, and both will likely get worse, but we don't know if these climate extremes will be expressed with more frequency or more intensity, or both.

It is reasonable to expect more precipitation, mostly during our existing wet seasons. More of our winter precipitation will fall as rain instead of snow, so storage of water in snowpacks will likely decrease (on average). We should expect milder winters, earlier melting of snow packs, earlier spring run-off, longer periods of summer low stream flow, and more drought. Importantly, earth's biogeochemical systems are complex and not at equilibrium. There are many feedbacks that lead to non-linear behaviour, so we should NOT expect climate changes to be slow and predictable. Small changes in CO_2 and global temperature can lead to large and/or rapid changes in climate and ecosystems. Accordingly, the rate of current and future global changes may be unprecedented, chaotic, and highly disruptive.

CLIMATE SYSTEM

The key to understanding global climate change is to first understand what global climate is, and how it operates. At the planetary scale, the global climate is regulated by how much energy the Earth receives from the Sun. However, the global climate is also affected by other flows of energy which take place within the climate system itself. This global climate system is made up of the atmosphere, the oceans, the ice sheets (cryosphere), living organisms (biosphere) and the soils, sediments and rocks (geosphere), which all affect, to a greater or less extent, the movement of heat around the Earth's surface.

The atmosphere plays a crucial role in the regulation of Earth's climate. The atmosphere is a mixture of different gases and aerosols (suspended liquid and solid particles) collectively known as air. Air consists mostly of nitrogen (78%) and oxygen (21%). However, despite their relative scarcity, the so-called greenhouse gases, including carbon dioxide and methane, have a dramatic effect on the amount of energy that is stored within the atmosphere, and consequently the Earth's climate.

These greenhouse gases trap heat within the lower atmosphere that is trying to escape to space, and in doing so, make the surface of the Earth hotter. This heat trapping is called the natural greenhouse effect, and keeps the Earth 33°C warmer than it would otherwise be. In the last 200 years, man-made emissions of greenhouse gases have enhanced the natural greenhouse effect, which may be causing global warming.

The atmosphere however, does not operate as an isolated system. Flows of energy take place between the atmosphere and the other parts of the climate system, most significantly the world's oceans. For example, ocean currents move heat from warm equatorial latitudes to colder polar latitudes. Heat is also transferred via moisture.

Water evaporating from the surface of the oceans stores heat which is subsequently released when the vapour condenses to form clouds and rain. The significance of the oceans is that they store a much greater quantity of heat than the atmosphere. The top 200 metres of the world's oceans store 30 times as much heat as the atmosphere. Therefore, flows of energy between the oceans and the atmosphere can have dramatic effects on the global climate.

The world's ice sheets, glaciers and sea ice, collectively known as the cryosphere, have a significant impact on the Earth's climate. The cryosphere includes Antarctica, the Arctic Ocean, Greenland, Northern Canada, Northern Siberia and most of the high mountain ranges throughout the world, where sub-zero temperatures persist throughout the year. Snow and ice, being white, reflect a lot of sunlight, instead of absorbing it. Without the cryosphere, more energy would be absorbed at the Earth's surface rather than reflected, and consequently the temperature of the atmosphere would be much higher.

All land plants make food from the photosynthesis of carbon dioxide and water in the presence of sunlight. Through this utilisation of carbon dioxide in the atmosphere, plants have the ability to regulate the global climate.

In the oceans, microscopic plankton utilise carbon dioxide dissolved in seawater for photosynthesis and the manufacture of their tiny carbonate shells. The oceans replace the utilised carbon dioxide by "sucking" down the gas from the atmosphere.

When the plankton die, their carbonate shells sink to the seafloor, effectively locking away the carbon dioxide from the atmosphere. Such a "biological pump" reduces by at least four-fold the atmospheric concentration of carbon dioxide, significantly weakening the Earth's natural greenhouse effect, and reducing the Earth's surface temperature.

VARIABILITY OF CLIMATE CHANGE

Variability

Variability — that is the outstanding characteristic of temperate-zone weather. Even the brief sea log just quoted shows how weather in the tropics is much the same from place to place and day to day; temperate-zone weather, conversely, is diverse and changeable. But had our voyage extended far enough northward, it would have reached another region of comparative weather sameness — cold sameness instead of warm sameness -the arctic zone.

The temperate zone can have no stable and typical weather of its own. To southward (in the northern hemisphere) is the steady, warm, moist climate of the tropics. To northward is the fairly steady, mostly cold, fairly dry climate of the arctic. (This picture is of course repeated, with reverse directions, in the southern hemisphere.) The temperate zone is a buffer region between tropic and arctic climates, a battleground on which the gigantic atmospheric forces of heat and moisture advance and retreat, an unstable compromise between irreconcilable extremes, swayed this way and that in its evanescent weather as tropic forces (warm air masses) or arctic forces (cold air masses) momentarily gain ascendancy.

Tropic Climate

Tropic wind-and-climate belts migrate somewhat, north and south, with the apparent north-south seasonal swing of the sun. To get within-the-tropics weather in its most typical form we must go well south of the Tropic of Cancer, nearly down to the equator — say to Panama in about lat. 8° N. In these latitudes there are only two main seasons — a winter dry season under the influence of the steady northeast trades, and a summer wet season brooded over by the fitful calms of the doldrums.

The outstanding features of tropical-marine climate are that the average temperature is practically constant from month to month (nor does it ever vary, by more than a few degrees, from day to day); and that the cloudiness and rainfall, always appreciable, reach high values during the wet-season doldrums invasion of summer and early fall.

Upper winds over Panama, at half-mile altitude, average northeast about 8 m.p.h. in the rainy season and north-northeast about 20 m.p.h. in the dry. At two miles altitude there is less seasonal change — the average wind only varies from east 10 m.p.h. (rainy) to cast-northeast 12 m.p.h. (dry). Upper-air characteristics over the Caribbean Sea, quite typical of tropical-marine climate in general, were studied during the 1939 hurricane season by a United States Weather Bureau expedition to Swan Island.

This four-mile-long islet, named after a nineteenth-century pirate, lies almost exactly in the middle of the Sea at about lat. 16° N., between Jamaica and British Honduras. Within four months, observers Rhamlow and Leech released some two hundred and fifty pilot balloons and one hundred and twenty radio-sounding balloons. They found that the trade wind extended well aloft, though with diminishing force, here as in Panama — prevailing easterlies all the way up to about two and a half miles altitude. But above this level, the upper winds are mostly out of the west and increase with altitude.

Surface temperatures at Swan Island, and all over the Caribbean area, average about +75° to 80° F. the year round. But at the base of the stratosphere, ten miles up in these latitudes, it is cold indeed — about — 105° to — 110° F. Thus in the thick tropical troposphere there is an average lapse rate of something like — 3½° F/1000 ft (about — 6° C/Km). Together with plenty of warmth and moisture in the lower levels, this lapse rate means that conditional instability must extend through considerable air layers, particularly in the vertical-sun-warmed rainy season of summer and fall — and it is this conditional instability, together with inexhaustible supplies of moisture from the warm ocean, that produces the daily torrential showers and thundershowers, the regular afternoon downpours and the moist, misty mornings which can make equatorial regions such a continual sweat-bath for ill-adapted white men. On or close to the ocean this deep-tropical weather is bearable, or perhaps even comfortable. But the steaming jungles inland, where temperatures can rise unchecked and where all breeze is choked off by the tangled uprearing of dense vegetation, are sometimes aptly described as 'green hells.' As Julian Duguid wrote '... a rich eternal garment of green, dappled with golden sunspots... a dense, fever-stricken thicket shimmering in the heat with a perpetual glassy haze....' However, by no means all the land areas within the tropics are covered with jungle — some of them are sandy and arid in the extreme, and doubly hot though not so muggy. (The word 'jungle' itself, originally, meant only 'waste land' in India.) And even in the fecund jungle areas of the tropics, such as Panama and northern South America, vegetation waxes lush and green with the rainy season, wanes sparse and tawny with the dry season.

There are also irregular year-to-year variations in moisture and prevailing winds that change the face of tropical nature. I got off a ship on the Colombian coast in August, 1939, and walked inland for a while, expecting to wade through wetseason green tangles of vines and trees peopled with tropic creatures — but the soil proved to be dry and sandy, the jungle a dried-up shrinkage of scrub, the creatures hidden away from the glaring vertical sun.

PHYSICAL EVIDENCE FOR CLIMATIC CHANGE

Evidence for climatic change is taken from a variety of sources that can be used to reconstruct past climates. Reasonably complete global records of surface temperature are available beginning from the mid-late 19th century. For earlier periods, most of the evidence is indirect—climatic changes are inferred from changes in proxies, indicators that reflect climate, such as vegetation, ice cores, dendrochronology, sea level change, and glacial geology.

Historical and Archaeological Evidence

Climate change in the recent past may be detected by corresponding changes in settlement and agricultural patterns. Archaeological evidence, oral history and historical documents can offer insights into past changes in the climate. Climate change effects have been linked to the collapse of various civilizations.

Glaciers

Glaciers are considered among the most sensitive indicators of climate change, advancing when climate cools and retreating when climate warms. Glaciers grow and shrink, both contributing to natural variability and amplifying externally forced changes.

A world glacier inventory has been compiled since the 1970s, initially based mainly on aerial photographs and maps but now relying more on satellites. This compilation tracks more than 100,000 glaciers covering a total area of approximately 240,000 km^2, and preliminary estimates indicate that the remaining ice cover is around 445,000 km^2.

The World Glacier Monitoring Service collects data annually on glacier retreat and glacier mass balance From this data, glaciers worldwide have been found to be shrinking significantly, with strong glacier retreats in the 1940s, stable or growing conditions during the 1920s and 1970s, and again retreating from the mid 1980s to present.

The most significant climate processes since the middle to late Pliocene (approximately 3 million years ago) are the glacial and interglacial cycles. The present interglacial period (the Holocene) has lasted about 11,700 years. Shaped by orbital variations, responses such as the rise and fall of continental ice sheets and significant sea-level changes helped create the climate. Other changes, including Heinrich events, Dansgaard–Oeschger events and the Younger Dryas, however, show how glacial variations may also influence climate without the orbital forcing.

Glaciers leave behind moraines that contain a wealth of material—including organic matter, quartz, and potassium that may be dated—recording the periods in which a glacier advanced and retreated. Similarly, by tephrochronological techniques, the lack of glacier cover can be identified by the presence of soil or volcanic tephra horizons whose date of deposit may also be ascertained.

Vegetation

A change in the type, distribution and coverage of vegetation may occur given a change in the climate; this much is obvious. In any given scenario, a mild change in climate may result in increased precipitation and warmth, resulting in improved plant growth and the subsequent sequestration of airborne CO_2. Larger, faster or more radical changes, however, may well result in vegetation stress, rapid plant loss and desertification in certain circumstances.

Ice Cores

Analysis of ice in a core drilled from a ice sheet such as the Antarctic ice sheet, can be used to show a link between temperature and global sea level variations.

The air trapped in bubbles in the ice can also reveal the CO_2 variations of the atmosphere from the distant past, well before modern environmental influences. The study of these ice cores has been a significant indicator of the changes in CO_2 over many millennia, and continues to provide valuable information about the differences between ancient and modern atmospheric conditions.

Dendroclimatology

Dendroclimatology is the analysis of tree ring growth patterns to determine past climate variations. Wide and thick rings indicate a fertile, well-watered growing period, whilst thin, narrow rings indicate a time of lower rainfall and less-than-ideal growing conditions.

Pollen Analysis

Palynology is the study of contemporary and fossil palynomorphs, including pollen. Palynology is used to infer the geographical distribution of plant species, which vary under different climate conditions. Different groups of plants have pollen with distinctive shapes and surface textures, and since the outer surface of pollen is composed of a very resilient material, they resist decay. Changes in the type of pollen found in different layers of sediment in lakes, bogs, or river deltas indicate changes in plant communities. These changes are often a sign of a changing climate. As an example, palynological studies have been used to track changing vegetation patterns throughout the Quaternary glaciations and especially since the last glacial maximum.

Insects

Remains of beetles are common in freshwater and land sediments. Different species of beetles tend to be found under different climatic conditions. Given the extensive lineage of beetles whose genetic makeup has not altered significantly over the millennia, knowledge of the present climatic range of the different species, and the age of the sediments in which remains are found, past climatic conditions may be inferred.

Sea Level Change

Global sea level change for much of the last century has generally been estimated using tide gauge measurements collated over long periods of time to give a long-term average. More recently, altimetre measurements — in combination with accurately determined satellite orbits — have provided an improved measurement of global sea level change. To measure sea levels prior to instrumental measurements, scientists have dated coral reefs that grow near the surface of the ocean, coastal sediments, marine terraces, ooids in limestones, and nearshore archaeological remains. The predominant dating methods used are uranium series and radiocarbon, with cosmogenic radionuclides being sometimes used to date terraces that have experienced relative sea level fall.

CLIMATE CHANGE FACTORS

Climate change is the result of a great many factors including the dynamic processes of the Earth itself, external forces including variations in sunlight intensity, and more recently by human activities. External factors that can shape climate are often called climate forcings and include such processes as variations in solar radiation, deviations in the Earth's orbit, and the level of greenhouse gas concentrations. There are a variety of climate change feedbacks that will either amplify or diminish the initial forcing.

Most forms of internal variability in the climate system can be recognized as a form of hysteresis, where the current state of climate does not immediately reflect the inputs. Because the Earth's climate system is so large, it moves slowly and has time-lags in its reaction to inputs. For example, a year of dry conditions may do no more than to cause lakes to shrink slightly or plains to dry marginally. In the following year however, these conditions may result in less rainfall, possibly leading to a drier year the next. When a critical point is reached after "*x*" number of years, the entire system may be altered inexorably. In this case, resulting in no rainfall at all. It is this hysteresis that has been mooted to be the possible progenitor of rapid and irreversible climate change.

Plate tectonics

On the longest time scales, plate tectonics will reposition continents, shape oceans, build and tear down mountains and generally serve to define the stage upon which climate exists. During the Carboniferous period, plate tectonics may have triggered the large-scale storage of Carbon and increased glaciation. More recently, plate motions have been implicated in the intensification of the present ice age when, approximately 3 million years ago, the North and South American plates collided to form the Isthmus of Panama and shut off direct mixing between the Atlantic and Pacific Oceans.

Solar output

The sun is the source of a large percentage of the heat energy input to the climate system. Lesser amounts of energy is provided by the gravitational pull of the Moon (manifested as tidal power), and geothermal energy. The energy output of the sun, which is converted to heat at the Earth's surface, is an integral part of the Earth's climate. Early in Earth's history, according to one theory, the sun was too cold to support liquid water at the Earth's surface, leading to what is known as the Faint young sun paradox. Over the coming millennia, the sun will continue to brighten and produce a correspondingly higher energy output; as it continues through what is known as its "main sequence", and the Earth's atmosphere will be affected accordingly.

On more contemporary time scales, there are also a variety of forms of solar variation, including the 11-year solar cycle and longer-term modulations. However, the 11-year sunspot cycle does not appear to manifest itself clearly in the climatological data. Solar intensity variations are considered to have been influential in triggering the Little Ice Age, and for some of the warming observed from 1900 to 1950. The cyclical nature of the sun's energy output is not yet fully understood; it differs from the very slow change that is happening within the sun as it ages and evolves, with some studies pointing toward solar radiation increases from cyclical sunspot activity affecting global warming.

Solar variations are changes in the amount of solar radiation emitted by the Sun. There are periodic components to these variations, the principal one being the 11-year solar cycle (or sunspot cycle), as well as aperiodic fluctuations. Solar activity has been measured via satellites during recent decades and through 'proxy' variables in prior times. Climate scientists are interested in understanding what, if any, effect variations in solar activity have on the Earth. Effects on the earth caused by solar activity are called "solar forcing".

The variations in total solar irradiance remained at or below the threshold of detectability until the satellite era, although the small fraction in ultra-violet wavelengths varies by a few percent. Total solar output is now measured to vary (over the last three 11-year sunspot cycles) by approximately 0.1% or about 1.3 W/m^2 peak-to-trough during the 11 year sunspot cycle. The amount of solar radiation received at the outer surface of Earth's atmosphere averages 1,366 watts per square meter (W/m^2). There are no direct measurements of the longer-term variation and interpretations of proxy measures of variations differ.

On the low side North et al. report results suggesting ~ 0.1% variation over the last 2,000 years. Others suggest the change has been ~ 0.2% increase in solar irradiance just since the 17th century. The combination of solar variation and volcanic effects are likely to have contributed to climate change, for example during the Maunder Minimum. Apart from solar brightness variations, more subtle solar magnetic activity influences on climate from cosmic rays or the Sun's ultraviolet radiation cannot be excluded although confirmation is not at hand since physical models for such effects are still too poorly developed.

The longest recorded aspect of solar variations is changes in sunspots. The first record of sunspots dates to around 800 BC in China and the oldest surviving drawing of a sunspot dates to 1128. In 1610, astronomers began using the telescope to make observations of sunspots and their motions. Initial study was focused on their nature and behavior. Although the physical aspects of sunspots were not identified until the 1900s, observations continued.

Study was hampered during the 1600s and 1700s due to the low number of sunspots during what is now recognized as an extended period of low solar activity, known as the Maunder Minimum. By the 1800s, there was a long enough record of sunspot numbers to infer periodic cycles in sunspot activity. In 1845, Princeton University professors Joseph Henry and Stephen Alexander observed the Sun with a thermopile and determined that sunspots emitted less radiation than surrounding areas of the Sun. The emission of higher than average amounts of radiation later were observed from the solar faculae.

Around 1900, researchers began to explore connections between solar variations and weather on Earth. Of particular note is the work of Charles Greeley Abbot. Abbot was assigned by the Smithsonian Astrophysical Observatory (SAO) to detect changes in the radiation of the Sun. His team had to begin by inventing instruments to measure solar radiation. Later, when Abbot was head of the SAO, it established a solar station at Calama, Chile to complement its data from Mount Wilson Observatory.

He detected 27 harmonic periods within the 273-month Hale cycles, including 7, 13, and 39 month patterns. He looked for connections to weather by means such as matching opposing solar trends during a month to opposing temperature and precipitation trends in cities. With the advent of dendrochronology, scientists such as Waldo S. Glock attempted to connect variation in tree growth to periodic solar variations in the extant record and infer long-term secular variability in the solar constant from similar variations in millennial-scale chronologies.

Statistical studies that correlate weather and climate with solar activity have been popular for centuries, dating back at least to 1801, when William Herschel noted an apparent connection between wheat prices and sunspot records. They now often involve high-density global datasets compiled from surface networks and weather satellite observations and/or the forcing of climate models with synthetic or observed solar variability to investigate the detailed processes by which the effects of solar variations propagate through the Earth's climate system.

MEASUREMENT OF CLIMATE ELEMENTS

Measurement of Temperature

Many surface air temperature records extend back to the middle part of the last century. The measurement of the surface air temperature is essentially the same now as it was then, using a mercury-in-glass thermometre, which can be calibrated accurately and used down to -39°C, the freezing point of mercury. For lower temperatures, mercury is usually substituted by alcohol. Maximum and minimum temperatures measured during specified time periods, usually 24 hours, provide useful information for the construction and analysis of temperature time series.

Analysis involves the calculation of averages and variances of the data and the identification, using various statistical techniques, of periodic variations, persistence and trends in the time series. Observations of temperature from surface oceans are also collected in order to construct time series.

In recent decades, much effort has also been directed towards the measurement of temperature at different levels in the atmosphere. There are now two methods of measuring temperatures at different altitudes: the conventional radiosonde network; and the microwave-sounding unit (MSU) on the TIROS-N series of satellites. The conventional network extends back to 1958 and the MSU data to 1979.

Temperature is a valuable climate element in climate observation because it directly provides a measure of the energy of the system under inspection. For example, a global average temperature† reveals information about the energy content of the Earth-atmosphere system. A higher temperature would indicate a larger energy content. Changes in temperature indicate changes in the energy balance, the causes of which were discussed. Variations in temperature are also subject to less variability than other elements such as rainfall and wind. In addition, statistical analysis of temperature time series is often less complex than that associated with other series. Perhaps most importantly of all, our own perception of the state of the climate are intimately linked to temperature.

Measurement of Rainfall

Rainfall is measured most simply by noting periodically how much has been collected in an exposed vessel since the time of the last observation. Care must be taken to avoid underestimating rainfall due to evaporation of the collected water and the effects of wind. Time series can be constructed and analysis performed in a similar manner to those of temperature. The measurement of global rainfall offers an indirect or qualitative assessment of the energy of the Earth-atmosphere system. Increased heat storage will increase the rate of evaporation from the oceans (due to higher surface temperatures). In turn, the enhanced levels of water vapour in the atmosphere will intensify global precipitation. Rainfall is, however, subject to significant temporal and spatial variability, and the occurrence of extremes, and consequently, analysis of time series is more complex.

Measurement of Humidity

The amount of water vapour in the air can be described in at least 5 ways, in terms of:

- The water-vapour pressure;
- The relative humidity;
- The absolute humidity
- The mixing ratio
- The dewpoint.

A full account of these definitions may be found in Linacre. The standard instrument for measuring humidity is a psychrometre. This is a pair of identical vertical thermometres, one of which has the bulb kept wet by means of a muslin moistened by a wick dipped in water. Evaporation from the wetted bulb lowers its temperature below the air temperature (measured by the dry bulb thermometre). The difference between the two measured values is used to calculate the air's water-vapour pressure, from which the other indices of humidity can be determined.

Measurement of Wind

Wind is usually measured by a cup anemometre which rotates about a vertical axis perpendicular to the direction of the wind. The exposure of wind instruments is important any obstruction close by will affect measurements. Wind direction is also measured by means of a vane, accurately balanced about a truly vertical axis, so that it does not settle in any particular direction during calm conditions.

Homogeneity

Non-climatic influences - inhomogeneities - can and do affect climatic observations. Any analyst using instrumental climate data must first assess the quality of the observations. A numerical series representing the variations of a climatological element is called homogeneous if the variations are caused only by fluctuations in weather and climate.

Leaving aside the misrecording of data, the most important causes of inhomogeneity are:

- Changes in instrument, exposure and measuring technique (for example, when more technologically advanced equipment is introduced);
- Changes in station location (i.e. when equipment is moved to a new site);
- Changes in observation times and methods used to calculate daily averages; and
- Changes in the station environment, particularly urbanisation (for example, the growth of a city around a pre-existing meteorological station).

When assessing the homogeneity of a climate record, there are three major sources of information: the variations evident in the record itself; the station history; and nearby station data. Visual examination and statistical analysis of the station record may reveal evidence of systematic changes or unusual behaviour which suggest inhomogeneity. For example, there may be a step-change in the mean, indicating a change in station location. A steady trend may indicate a progressive change in the station environment, such as urbanisation. An extreme value may be due to a typing error. Often these inhomogeneities may be difficult to detect and other evidence is needed to confirm their presence.

One source of evidence is the station history, referred to as metadata. The station history should include details of any changes of location of the station, changes in instrumentation or changes in the timing and nature of observation. Very often, though, actual correction factors to observation data containing known inhomogeneities will be difficult to calculate, and in these cases, the record may have to be rejected.

The third approach to homogenisation involves empirical comparisons between stations close to each other. Over time scales of interest in climate change studies, nearby stations (i.e. within 10km of each other) should be subject to similar changes in monthly, seasonal and annual climate. The only differences should be random. Any sign of systematic behaviour in the differences (e.g. a trend or step-change) would suggest the presence of inhomogeneities. In light of the foregoing discussion on homogeneity, careful attention has to be paid to eliminating sources of non-climatic error when constructing large-scale record, such as the global surface air temperature time series. This, and similar records including sea surface temperatures, rely on the collection of millions of individual observations from a huge network made up of thousands of climate stations. A number of these have been reviewed by Jones*et al.*. The effects of urbanisation (the artificial warming associated with the growth of towns and cities around monitoring sites) were considered to be the greatest source of inhomogeneity, but even this, it was concluded, accounts for at most a 0.05°C warming (or 10% of the observed warming) over the last 100 years. Conrad and Pollak (1962), Folland *et al.* (1990), and Jones *et al.* provide useful references investigating the problems of homogeneity and data reliability of instrumental records of climate data.

Statistical Analysis of Instrumental Records

Once climate data has been collected and corrected for inhomogeneities, it will need to be analysed. The aim of any statistical analysis is to identify systematic behaviour in a data set and hence improve understanding of the processes at work to compliment the theory. Statistical analysis is a search for a signal in the data that can be distinguished from the background noise In climate change research that signal will be a periodic variation, a quasi-periodic variation, a trend, persistence or extreme events in the climate element under analysis.

Before undertaking a statistical analysis of a climate record, a number of questions about the task in hand should be asked.

1. What is the purpose of the analysis?

In its simplest form, the statistical analysis may be:

- Descriptive; or
- Investigative.

Descriptive analyses set out solely to document particular aspects of the variations present in the data set (the signal). Indices calculated will include the mean and variance (or standard deviation). The occurrence of extreme events, cycles and trends will also be noted. Significance testing is crucial to this category of analysis. Significance testing establishes whether or not the variation under consideration is different from what one would expect to arise in a random time series.

Investigative analyses set out to test a pre-defined hypothesis. The hypothesis should *a priori* have a sound physical basis. "Does the time series contain an El Niño cycle?" would be an example of a hypothesis that could be investigated.

2. What is the most appropriate data set to use?

Any data set used for a statistical analysis should be:

- Representative of the relevant physical processes;
- Sufficient in quantity to support the statistical method(s) to be used; and

- Accurate and reliable (homogeneous).

To investigate the impact of El Niño† on drought in eastern Australia, it is necessary to firstly identify a representative indicator of El Niño, such as sea surface temperatures in the SE Pacific. Secondly, a reliable indicator of drought in eastern Australia is required, for example, rainfall. The data set would need to be of sufficient duration to permit the testing for a relationship on the time scale under consideration, i.e. does El Niño cause drought in eastern Australia? Since El Niño has a quasi-periodicity of 2 to 5 years, then a data set of length at least 7 to 10 times this duration (i.e. up to 50 years) is required to have confidence in the statistical methods. To investigate longer-term trends, the data requirement becomes more stringent.

3. What is the most appropriate technique to use and how should it be applied?

Often it will be clear as to which statistical method of analysis is required. However, its application may be less straightforward. The nature of the data may determine whether or not a particular technique is valid (or, at least the way in which the technique is applied). For example, if the data are not normally distributed † then this may invalidate assumptions on which the technique is based. What ever technique is used, it goes without saying that testing statistical significance must be a critical concern.

Barry and Perry offer a detailed introduction of the mathematical aspects of statistical analysis, with many useful examples. Other useful references are provided by Gani and Godske. Before concluding, however, a couple of points need illustrating. Firstly, much of what has been said about statistical analysis of instrumental records applies equally well to the study of palaeoclimatology, and to the reconstruction of past climates from proxy data. Secondly, the statistical analysis of climate data serves to compliment and support theories developed to explain the causes (and effects) of climate change. Statistical associations do not prove cause and effect for they are solely based upon the laws of probability. When analysing and interpreting climate data in the effort to aid understanding of the causes of climate change, it is necessary to bare this in mind.

Historical Records

Historical records have been used to reconstruct climates dating back several thousand years (i.e. for most of the Holocene). Historical proxy data can be grouped into three major categories. First, there are observations of weather phenomena *per se*, for example the frequency and timing of frosts or the occurrence of snowfall. Secondly, there are records of weather-dependent natural or environmental phenomena, termed parameteorological phenomena, such as droughts and floods. Finally, there are phenological records of weather-dependent biological phenomena, such as the flowering of trees, or the migration of birds.

Major sources of historical palaeoclimate information include: ancient inscriptions; annals and chronicles; government records; estate records; maritime and commercial records; diaries and correspondence; scientific or quasi-scientific writings; and fragmented early instrumental records.

There are a number of major difficulties in using this kind of information. First, it is necessary to determine exactly what the author meant in describing the particular event. How severe was the "severe" frost? What precisely does the term drought refer to? Content analysis - a standard historical technique - has been used to assess, in quantitative terms, the meaning of key climatological phrases in historical accounts. This approach involves assessment of the frequency of use made of certain words or phrases by a particular author. Nevertheless, the subjectivity of any personal account has to be carefully considered. Very often, the record was not kept for the benefit of the future reader, but to serve some independent purpose. During much of the dynastic era in China, for example, records of droughts and floods would be kept in order to gain tax exemptions at times of climatic adversity.

Secondly, the reliability of the account has to be assessed. It is necessary to determine whether or not the author had first-hand evidence of the meteorological events. Thirdly, it is necessary to date and interpret the information accurately. The representativeness of the account has to be assessed. Was the event a localised occurrence or can its spatial extent be defined by reference to other sources of information? What was the duration of the event? a day? a month? a year? Finally, the data must, as with all proxy records, be calibrated against recent observations and cross-referenced with instrumental data This might be achieved by a construction of indices (e.g. the number of reports of frost per winter) which can be statistically related to analogous information derived from instrumental records.

Ice Cores

As snow and ice accumulates on polar and alpine ice caps and sheets, it lays down a record of the environmental conditions at the time of its formation. Information concerning these conditions can be extracted from ice and snow that has survived the summer melt by physical and chemical means. When melting does occur, the refreezing of meltwater can provide a measure of the summer conditions.

Palaeoclimate information has been obtained from ice cores by three main approaches.

These involve the analysis of:

- Stable isotopes of water;
- Dissolved and particulate matter in the firn † and ice; and
- The physical characteristics of the firn and ice, and of air bubbles trapped in the ice. Each approach has also provided a means of dating the ice at particular depths in the ice core.

Stable isotope analysis

The basis for palaeoclimatic interpretations of variations in the stable isotope† content of water molecules is that the vapour pressure of $H2^{16}O$ is higher than that of $H2^{18}O$. Evaporation from a water body thus results in a vapour which is poorer in ^{18}O than the initial water; conversely, the remaining water is enriched in ^{18}O. During condensation, the lower vapour pressure of the $H2^{18}O$ ensures that it passes more readily into the liquid state than water vapour made up of the lighter oxygen isotope. During the poleward transportation of water vapour, such isotope fractionation continues this preferential removal of the heavier isotope, leaving the water vapour increasingly depleted in $H2^{18}O$. Because condensation is the result of cooling, the greater the fall in temperature, the lower the heavy isotope concentration will be. Isotope concentration in the condensate can thus be considered as a function of the temperature at which condensation occurs. Water from polar snow will thus be found to be most depleted in $H2^{18}O$.

This temperature dependency allows the oxygen isotope content of a ice core to provide a proxy climate record. The relative proportions of ^{16}O and ^{18}O in an ice core are expressed in terms of departures, $\delta^{18}O$, from the Standard Mean Ocean Water (SMOW) standard, such that:

$$\frac{\delta^{18}O = (^{18}O/^{16}O)_{sample} - (^{18}O/^{16}O)_{SMOW} \times 10^{3}\%}{\left(^{18}O/^{16}O\right)_{SMOW}}$$

All measurements are made using a mass spectrometre and results are normally accurate to within 0.1‰ (parts per mille). A $\delta^{18}O$ value of -10‰ indicates a sample with an $^{18}O/^{16}O$ ratio 1% or 10‰ less than SMOW. For most palaeoclimate reconstructions, typical values for $\delta^{18}O$ obtained from ice cores range between -10 and -60‰.

Similar palaeoclimatic studies can be carried out using isotopes of hydrogen (^{1}H and ^{2}H (Deuterium)), but these are rarer in nature and the laboratory techniques involved are more complex.

Physical and chemical characteristics of ice cores

The occurrence of melt features in the upper layers of ice cores are of particular palaeoclimatic significance. Such features include horizontal ice lenses and vertical ice glands which have resulted from the refreezing of percolating water. They can be identified by their deficiency in air bubbles.

The relative frequency of melt phenomena may be interpreted as an index of maximum summer temperatures or of summer warmth in general. Other physical features of ices cores which offer information to the palaeoclimatologist include variations in crystal size, air bubble fabric and crystallographic axis orientation.

Another important component of ice cores which is of palaeoclimatic significance is the atmospheric gas content, as the air pores are closed off during the densification of firn to ice. Considerable research effort has been devoted to the analysis of carbon dioxide concentrations of air bubbles trapped in ice cores. It will be seen that variations in atmospheric carbon dioxide may have played an important role in the glacial-interglacial climatic variations during the Quaternary.

Finally, variations of particulate matter, particularly calcium, aluminium, silicon and certain atmospheric aerosols can also be used as proxy palaeoclimatic indicators.

Dating ice cores

One of the biggest problems in any ice core study is determining the age-depth relationship. Many different approaches have been used and it is now clear that fairly accurate time scales can be developed for the last 10,000 years. Prior to that, there is increasing uncertainty about ice age. The problem lies with the fact that the age-depth is highly exponential, and ice flow models are often needed to determine the ages of the deepest sections of ice cores.

For example, the upper 1000m of a core may represent 50,000 years, whilst the next 50m may span another 100,000 year time period, due to the severe compaction, deformation and flow of the ice sheet in question.

Radio isotope dating†, using ^{210}Pb (lead), ^{32}Si (silicon), ^{39}Ar (argon) and ^{14}C (carbon) have all been used with varying degrees of success, over different time scales, to determine the age of ice cores.

Certain components of ice cores may reveal quite distinct seasonal variations which enable annual layers to be identified, providing accurate time scales for the last few thousand years.

Such seasonal variations may be found in $\delta^{18}O$ values, trace elements and microparticles (Hammer *et al.*, 1978).

Where characteristic layers of known ages can be detected, these provide valuable chronostratigraphic markers against which other dating methods can be verified. So-called reference horizons have resulted from major explosive volcanic eruptions. These inject large quantities of dust and gases (principally sulphur dioxide) into the atmosphere, where they are globally dispersed. The gases are converted into aerosols (principally of sulphuric acid) before being washed out in precipitation.

Hence, after major eruptions, the acidity of snowfall increases significantly above background levels. By identifying highly acidic layers (using electrical conductivity) resulting from eruptions of known age, an excellent means of checking seasonally based chronologies is available.

Dendroclimatology

The study of the annual growth of trees and the consequent assembling of long, continuous chronologies for use in dating wood is called dendrochronology. The study of the relationships between annual tree growth and climate is called dendroclimatology. Dendroclimatology offers a high resolution (annual) form of palaeoclimate reconstruction for most of the Holocene.

The annual growth of a tree is the net result of many complex and interrelated biochemical processes. Trees interact directly with the microenvironment of the leaf and the root surfaces. The fact that there exists a relationship between these extremely localised conditions and larger scale climatic parametres offers the potential for extracting some measure of the overall influence of climate on growth from year to year. Growth may be affected by many aspects of the microclimate: sunshine, precipitation, temperature, wind speed and humidity. Besides these, there are other non-climatic factors that may exert an influence, such as competition, defoliators and soil nutrient characteristics.

There are several subfields of dendroclimatology associated with the processing and interpretation of different tree-growth variables. Such variables include tree-ring width (the most commonly exploited information source, e.g. Briffa and Schweingruber,), densitometric parametres and chemical or isotopic variables.

A cross section of most temperate forest tree trunks † will reveal an alternation of lighter and darker bands, each of which is usually continuous around the tree circumference. These are seasonal growth increments produced by meristematic tissues in the cambium of the tree. Each seasonal increment consists of a couplet of earlywood (a light growth band from the early part of the growing season) and denser latewood (a dark band produced towards the end of the growing season), and collectively they make up the tree ring. The mean width of the tree ring is a function of many variables, including the tree species, tree age, soil nutrient availability, and a whole host of climatic factors.

The problem facing the dendroclimatologist is to extract whatever climatic signal† is available in the tree-ring data from the remaining background "noise".

Whenever tree growth is limited directly or indirectly by some climate variable, and that limitation can be quantified and dated, dendroclimatology can be used to reconstruct some information about past environmental conditions.

Only for trees growing near the extremities of their ecological amplitude†, where they may be subject to considerable climatic stresses, is it likely that climate will be a limiting factor. Commonly two types of climatic stress are recognised, moisture stress and temperature stress.

Trees growing in semi-arid regions are frequently limited by the availability of water, and dendroclimatic indicators primarily reflect this variable. Trees growing near the latitudinal or altitudinal treeline are mainly under growth limitations imposed by temperature; hence dendroclimatic indicators in such trees contain strong temperature signals.

Furthermore, climatic conditions prior to the growth period may precondition physiological processes within the tree and hence strongly influence subsequent growth. Consequently, strong serial correlation or autocorrelation may establish itself in the tree-ring record.

A specific tree ring will contain information not just about the climate conditions of the growth year but information about the months and years preceding it.

Several assumptions underlie the production of quantitative dendroclimatic reconstructions. First, the physical and biological processes which link toady's environment with today's variations in tree growth must have been in operation in the past. This is the principle of uniformitarianism.

Second, the climate conditions which produce anomalies in tree-growth patterns in the past must have their analogue during the calibration period.

Third, climate is continuous over areas adjacent to the domain of the tree-ring network, enabling the development of a statistical transfer function relating growth in the network to climate variability inside and outside of it. Finally, it is assumed that the systematic relationship between climate as a limiting factor and the biological response can be approximated as a linear mathematical expression. Fritts provides a more exhaustive review of the assumptions involved in the use of dendroclimatology.

Bradley gives a full account of the methods (1 to 6 above) of palaeoclimate reconstruction from tree-ring analysis. This approach may be applied to all the climate-dependent tree-growth variables, specifically tree-ring width, but also wood density and isotopic measurements.

The latewood of a tree ring is much denser than the earlywood and interannual variations contain a strong climatic signal. Density variations are particularly valuable in dendroclimatology because they to not change significantly with tree age, and the process of standardisation (removal of growth function) can therefore be avoided.

The use of isotopic measurements in dendroclimatology also avoids the need for a standardisation process. The basic premise of isotope dendroclimatology is that since $^{18}O/^{16}O$ and D/H (deuterium/hydrogen) variations in meteoric (atmospheric) waters are a function of temperature, tree growth which records such isotope variations should preserve a record of past temperature fluctuations. Unfortunately, isotope fractionation effects within the tree, which are themselves temperature dependent, will create problems associated with this technique.

CLIMATE CHANGE MITIGATION

Climate change mitigation is action to decrease the intensity of radiative forcing in order to reduce the potential effects of global warming. Mitigation is distinguished from adaptation to global warming, which involves acting to tolerate the effects of global warming. Most often, climate change mitigation scenarios involve reductions in the concentrations of greenhouse gases, either by reducing their sources or by increasing their sinks.

Scientific consensus on global warming, together with the precautionary principle and the fear of abrupt climate change is leading to increased effort to develop new technologies and sciences and carefully manage others in an attempt to mitigate global warming. Most means of mitigation appear effective only for preventing further warming, not at reversing existing warming. The Stern Review identifies several ways of mitigating climate change. These include reducing demand for emissions-intensive goods and services, increasing efficiency gains, increasing use and development of low-carbon technologies, and reducing fossil fuel emissions.

The energy policy of the European Union has set a target of limiting the global temperature rise to 2 °C compared to preindustrial levels, of which 0.8 °C has already taken place and another 0.5–0.7 °C is already committed. The 2 °C rise is typically associated in climate models with a carbon dioxide-equivalent concentration of 400–500 ppm by volume; the current level of carbon dioxide alone is 383 ppm by volume, and rising at 2 ppm annually. Hence, to avoid a very likely breach of the 2 °C target, CO_2 levels would have to be stabilised very soon; this is generally regarded as unlikely, based on current programmes in place to date. The importance of change is showed by the fact that world economic energy efficiency is presently improving at only half the rate of world economic growth.

Bibliography

Barrows, M.: *A Survey of the Intestinal Parasites of the Primates in Budongo Forest,* Uganda. Glasgow University, 1996.

Bethell, E.: *Vigilance in Foraging Chimpanzees.* University College, London University, 1998.

Biswas, A., and Cline, S.: *Global Warming: Impacts on Water and Food Security,* Springer, Heidelberg, 2010.

Burdak, L.R.: *Recent Advances in Desert Afforestation and Global Warming,* F.R.I., Dehra dun, 1982.

Burroughs, W.J.: *The Climate Revealed,* Cambridge University Press, Cambridge, 1999.

Carter, T. R. and Konijn, N. T. : . *The Impact of Climatic Variations on Agriculture.* Dordrecht, Netherlands: Kluwer, 1988.

Chouhan, T.S.: *Desertification in the World and Its Control,* Scientific Publishers, Delhi, 1992.

Colin M. *The Mountain People.* New York: Simon and Schuster, 1972.

Dasmann, R.F.: *Environmental Conservation,* New York, Wiley, 1972.

Ferentinos L.: *Proceeding of the Sustainable Taro Culture for the Pacific Conference,* Honolulu, HITAHR, 1993.

Geist, Helmut: *The Causes and Progression of Desertification,* Ashgate Publishing, Delhi, 2005.

Hobbie, J.E.: *Micro-organisms in Action: Concepts and Applications in Microbial Ecology,* Blackwell Scientific Publications, Oxford, 1988.

Jameson, J. L.: *Principles of Molecular Medicine,* Totowa, Humana Press, 1998.

Karliner, Joshua: *The Wmx Corporation, Hazardous Waste, and Global Strategies For Environmental Justice,* San Francisco, Political Ecology Group, 1994.

Konijn, N. T. : . *The Impact of Climatic Variations on Agriculture.* Dordrecht, Netherlands: Kluwer, 1988.

Lakshmi Narasaiah: *Irrigation Management and Globalisation,* Discovery, 2006.

Magalhaes, A. R. *Impacts of Climatic Variations and Sustainable Development in Semi-arid Regions.* Proceedings of International Conference. ICID. Fortaleza, Brazil, 1992.

Narasimha Rao, P.: *Irrigation Development: Issues and Challenges,* Discovery, 2007.

Oliver, John E.: *Encyclopedia of World Climatology,* Springer, Delhi, 2005.

Parry M.L.: *Climatic Change, Agriculture and Settlements,* Dawson Folkestone UK, 1978.

Reynolds, F.: *Chimpanzees of the Budongo Forest,* New York, Rinehart and Winston, 1965.

Sheldon, J.K.: *Practical Environmental Bioremediation,* Lewis Publishers, Boca Raton, 1992.

Switzer, R.L.: *Experimental Biodiversity,* New York, W.H. Freeman and Company, 1977.

Tucker, Mary Evelyn, and John Grim: *Worldviews and Ecology: Religion, Philosophy, and the Environment.* Orbis Books, Maryknoll, N.Y. 1994.

Walter, B., L. Arkin : *Sustainable Cities, Concepts and Strategies for Eco-city Development,* Los Angeles, CA, EHM Eco-Home Media, 1992.

Webb, R. : *Floods, Droughts and Climate Change,* University of Arizona Press, NY, 2001.

White, G.F.: *Natural Hazards: Local, National, Global,* Oxford University Press, New York, 1974.

Index